Mystery Tusk

Searching for Elephants in the Maine Woods

by Gary Hoyle

Illustrations by the Author

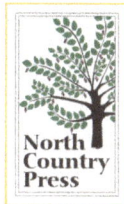

North Country Press

ISBN 978-1-943424-73-3

LCCN 2022937821

Cover art by Gary Hoyle

North Country Press
Unity, Maine

Dedicated to the memory of the keen observers:
William Littlejohn, Leonard Cash, and Christopher Packard

And to my friend and colleague on the Mammoth Project:
the late Harold Borns, Jr.

Foreword

"Hello, this is Gary Hoyle," I said into the telephone. "I'm the Curator of Natural History at the Maine State Museum. I don't know if I have the right Littlejohn."

"Is it about the tusk?" Janice Littlejohn replied.

That phone call in July 1990 put me on a path to the epicenter of the largest paleontological excavation in Maine history, involving roughly a hundred volunteers and professional researchers. But the origin of that path began many decades before.

Table of Contents

1. How It All Began

On the morning of August 31, 1959, Leonard Cash moved his cable-driven power shovel farther up the gulley from a dry hole he had dug the previous day to a spot where a dowser claimed he would find water. James Littlejohn had hired Cash to dig a pond for watering nursery stock at his farm in North Scarborough, Maine, and the logical location was this tiny tributary of the Nonesuch River, which bordered one of Littlejohn's fields. That day Cash had an assistant working with him, James' son, William.

Author's Illustration

When Cash's shovel tore into the wall of the gulley, Littlejohn cleared away the spoils with a small bulldozer. Cash quickly dug through layers of sand, but about three feet down, a mass of dense, brown clay hindered his progress. Each raking of the shovel's teeth brought up only crumbs in the bucket until he cut through and sliced into a plastic, gray clay that easily peeled away in layers.

As the shovel lifted one of the layers, Cash saw a half-moon trough impressed in the clay hanging from his bucket. A flash of white in the hole made him stop the machine. How could a root be that deep in the clay? As he jumped off his power shovel, Littlejohn ran to his side. They slid across the clay and stopped at the tubular object. It wasn't a root. Lying with half its profile exposed above a veneer of sand, a six- foot ivory tusk glistened in the sunlight.

Both men tried lifting it from the clay, but a crust of compacted sand prevented them from working their fingers under the tusk. Leonard Cash jumped on his power shovel and lowered the bucket into the hole. The teeth of the shovel crept behind the tip of the tusk and cut through the sand layer like a knife through soft butter. But just as the tusk began rising, it shattered, spraying ivory splinters across the clay. Over a foot of the ivory tip broke away.

Fearing more damage if they continued, work halted. William Littlejohn called Emma Downs, a local reporter, who in turn called the Portland Museum of Natural History. In a short time Christopher Packard, the Portland Museum's director, arrived at the site.[1]

Finding an abundance of seashells in the sand layer associated with the tusk, Packard speculated that the tusk might be from an ancient walrus. A few remains of them had been found at various locations in Maine. However, he felt uncomfortable with his hypothesis because the large size and curvature of the tusk did not conform to characteristics of known walrus species. When he got home, he called Dr. Bryan Patterson, a vertebrate paleontologist at Harvard University.

Packard gave Patterson a full account of the tusk's discovery along with a detailed description of its *in situ* environment. When he described the length and curvature of the tusk, Patterson took great interest. It came from no walrus. It had to be from one of the ancient elephants. But could it be from a circus elephant? Not in a layer of clay with sand and shells over it.

Packard was ecstatic. The following day he told a newspaper reporter of his hope to find a complete skeleton of the prehistoric beast so it could be displayed at the Portland museum. But the clay posed a problem.

"'It's pretty messy and I'm afraid we're the ones who are going to be dug out by some future generation.'" [2]

To make matters worse, rain crippled progress over the next few days. Then on Friday, September 4, the weather cleared, and Packard arrived at the site with his army of diggers. But they had to wait for another problem to be solved. Four feet of water filled the thirty-foot long hole. [3]

Early that afternoon James Littlejohn pumped out the pond with his irrigation system, and people began probing and shoveling. By the end of the day they had found nothing of the animal. Chances looked bleak. Geologists had discovered a gravel layer two to three feet below the clay where conditions were too acidic for bone to survive long. The target now centered on a narrow clay band.

At that point Packard believed that not much of the animal remained buried in the clay, but he still held hope of finding one element.

"Dr. Bryan Patterson at Harvard told me that the molar is the thing. From it, he can tell everything we want to know." [4]

But he never found a tooth.

2. Got Bones?

In 1989 just as I finished a display case of realistic artificial plants for the Maine State Museum's exhibit hall "12,000 Years in Maine," the curator of that hall, our chief archaeologist Bruce Bourque, told me that he needed some large Ice Age fossils for an introductory case. Faced with the paucity of specimens from Maine, Bruce became concerned that we would have to exhibit fossils from elsewhere. For many museums specimen locality would not be an important exhibit criterion, but over the years we had prided ourselves in exhibiting primarily Maine objects, and since we had received a national award for our last exhibit hall, "Made in Maine," saturated with Maine artifacts,[1] it seemed almost incomprehensible that we would resort to displaying out-of-state fossils as an introduction to a hall already half full of Maine archaeological objects.

A few good Maine Ice Age fossils could be gathering dust in some scientist's office, but with his commitments to exhibits, research, teaching, and publishing, Bruce had no time to look for them. But I did. And the first person I called was my friend Woody Thompson.

Dr. Woodrow Thompson worked as a physical geologist at the Maine Geological Survey where his interests ranged from glaciers to rocks and minerals. Over the years he had proven invaluable to me in procuring significant Maine gem quality minerals for the museum. This time Woody put on his paleontological hat.

Woody listed off a few seal bones and walrus skulls that had been found in clay beds around the state, most of which were in the process of being acquired through Bruce for our museum.

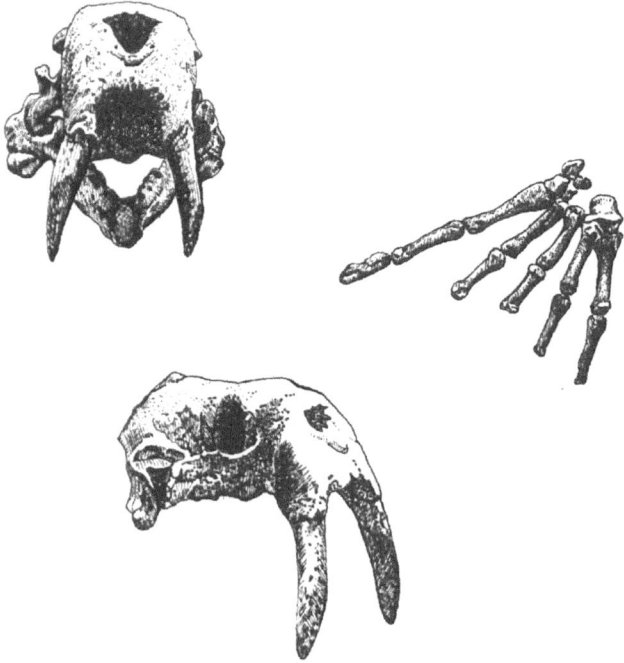

Walrus skulls (male – top, female – bottom) and bones of a seal flipper recovered from glacial clay. (Author's illustration)

Then, in what seemed like an afterthought, he mentioned an ancient elephant tusk hanging on a wall outside the office of Dr. Harold Borns in the Institute of Quaternary Studies at the University of Maine. It may have come from Scarborough, Maine, but that remained controversial. If interested, I should contact Dr. Borns.

3. A Provenance Dilemma

When I first went to the Institute of Quaternary Studies, I found it in the upper stories of Boardman Hall on the university's Orono campus. Today, it thrives in a larger more modern building known as the Edward Bryand Global Science Center at the other end of campus. Its name has been changed to the Climate Change Institute, which more broadly describes the Institute's purpose. Studies focus worldwide on the Quaternary Period: the time from the beginning of the last Ice Age to the present.

Entering the Institute's outer office, a childhood thrill raced through me. A ten-foot tusk, markedly curved, weathered and mineral stained, loomed from the wall: a classic symbol of the Ice Age.

HM9258. Hudson Museum, University of Maine. Photo Credit: Gretchen Faulkner.

I almost prayed the tusk came from Scarborough. At the same time, I became apprehensive about meeting Dr. Borns because I felt that I might be infringing on his valuable time. Not only did he head the Institute, he was well known for his work in Antarctica and for the glacier named after him. But his solid handshake and greeting of, "Hi, I'm Hal," put me at ease.

In his office Hal spoke about finding the tusk in the northern part of the state at the University of Maine at Presque Isle, in part of the Portland Society of Natural History collection. That society had been the owner of the Portland Museum of Natural History in Portland, Maine, where in 1971 a heart-wrenching event took place for natural history lovers. The museum was razed for a parking lot. By then, that society had been incorporated into the Maine Audubon Society due to falling membership and reduced financial support.

Gone went a respected institution established in 1843 of which notable members included the poet Henry Wadsworth Longfellow, the North Pole discoverer Commodore Robert Peary, and the first director of Harvard's Botanical Museum George L. Goodale, who commissioned the creation of the famed "glass-flower" collection. The Society's library and natural history collections were once on a par with those at Harvard, Yale, the Boston Society of Natural History, and the Essex Museum (Salem, MA). [1]

The city condemned the museum building while officials were rushing to find homes for the collections. That action resulted in a failure to record all the acquisition sites. Today many of these locations are unknown. However, thanks to the efforts of staff geology instructor William Forbes, the University of Maine at Presque Isle became one of those known repositories.

Among the many boxes Forbes brought to Presque Isle was a large crate containing two tusks, one labeled Alaska, the other Scarborough, Maine. Over a decade passed before anyone opened the crate. By then, the labels had peeled off and fallen to the bottom. Neither Bill Forbes nor anyone else associated with the move knew which tusk came from Scarborough.

When Hal finished his story, he speculated that the Scarborough tusk had been collected in the 1800s. Fire destroyed the Portland Society's collection on two different occasions. During the first conflagration in 1854, one of the priceless objects lost had been the only Maine timber rattlesnake in a natural history collection (a species now locally extinct). The last fire in 1866 put a 34-year limit on our 19th century search. But that proved to be a lot of records, and Hal and I both had pressing projects to finish. We agreed to research the problem when we could.

With Hal's permission, I borrowed the tusk to study and photograph at the Maine State Museum. I also wanted to know about the other tusk that had been found in the crate. That specimen, Hal said, was smaller and in poor condition. Having been broken in two, it shed ivory chips and retained multiple sooty gray stains. Obviously Hal chose the better tusk for display.

At the museum I used a magnifier to examine every square inch of the tusk in the hope of finding a clue that would nail down its provenance. Having never heard of any prehistoric elephant being found in Maine, Hal's speculation that the tusk had been found in the 19th century seemed likely. It also seemed likely that such a discovery would have been mentioned in a Portland newspaper.

I called Herbert Adams who at that time served as a member of the Maine House of Representatives. Herb, who had an uncanny resemblance to the late actor Gene Wilder, prided himself as a Portland historian and often walked next door to the museum, when taking a break from his legislative duties. He worked also as a reporter for the *Portland Evening Express*, which had begun merging with the *Portland Press Herald*. He could not recall any historical references to ancient elephants in Maine, but his newspaper had recently indexed its files, and he gladly offered to have a look.

A few weeks later, to my surprise, Hal Borns called to tell me that he had just gotten off the phone with Herb Adams. Herb had found a brief statement in an early September 1959 *Portland Evening Express* edition that mentioned two workers finding a mastodon tusk in Scarborough, Maine.

4. Identity Crisis

The American mastodon, known scientifically as *Mammut americanum*, evolved as a distant cousin to modern day elephants. During the Ice Age it ranged across the continent north to Alaska, parts of Canada and New England, and south to Mexico and Florida. Masses of well-preserved plant material have been found below the rib cages of some excavated specimens, indicating a preferred diet of browse.

The conical ridges on the crown of their molar teeth functioned specifically for processing such a diet. These projections have given the mastodon its name meaning breast tooth. This evidence along with its associated geology suggests that wet, open woodlands became prime mastodon habitat.

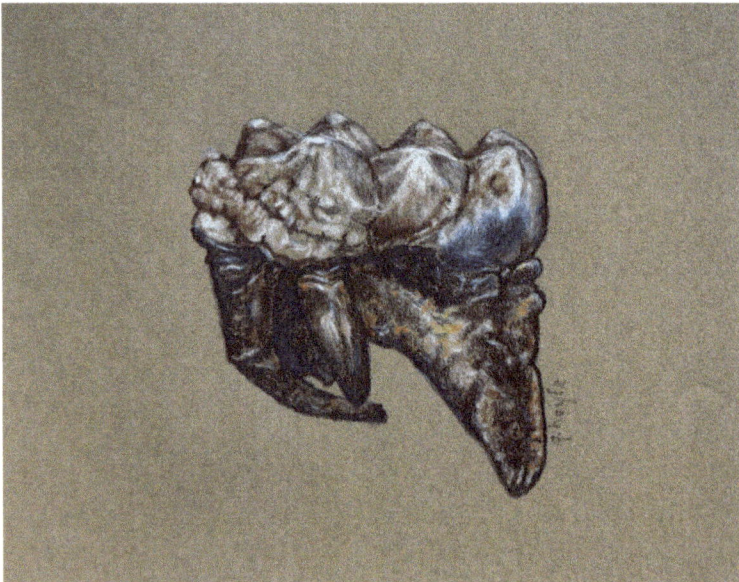

Mastodon tooth. (Author's illustration)

The animal's overall stature appeared similar to that of the Asian elephant except the skull grew long and set low, and hair covered its body. Both sexes sported tusks, which grew longer and heavier in males. Generally, the tusk curvature took on a gentle, spiral sweep typical of that found in modern elephants.

Mastodon. (Author's illustration)

In contrast to the mastodon, the iconic symbol of the Ice Age, the woolly mammoth [*Mammuthus primagenius*], normally grew more tightly spiraled tusks. Like the mastodon, tusks developed in both sexes. In some males they extended over twelve feet in length. Those of females grew less than half that. Despite the name mammoth, this close relative to the Asian elephant stood slightly shorter than the African species, but its girth likely appeared larger due to a nearly four-inch layer of subcutaneous fat and a dense three-inch covering of under-wool beneath its longer guard hairs.

Wooly Mammoth. (Author's illustration)

The overall body shape of the woolly mammoth has been documented by over 400 prehistoric cave paintings and frozen remains found in the permafrost of Siberia and Alaska. Behind the high, domed head, a hump rose over the shoulders, tapering to the hips. The long guard hairs, covering wool on the flanks and underside, created a "skirt" of protection against the brutally cold winds that raged across the grasslands south of the ice sheet, where mammoths foraged. The ribbed grinding surface of the high crowned molar teeth functioned like a millstone to process forbs (flowering plants such as yarrow, mums, tansies, and sagebrush) and grasses.

The woolly mammoth evolved in western Siberia from a larger species that ranged across Europe and Siberia. In the late Ice Age it crossed the Bering land bridge into North America and expanded its range south into northern states, where it comingled with another larger mammoth species that had crossed the land bridge over a million years earlier.

*　　*　　*　　*　　*　　*

As important as all this information became to me, at that moment my focus settled on the pivotal blind spot of my ignorance. I lacked the valuable experience of handling the tusks of these two species. The museum had no examples from mammoth, mastodon, or even modern day elephant to help me evaluate Hal's tusk. I knew from my zoological training that variability occurred in all species. Only an experienced eye could decipher what species I dealt with and whether it had come from Alaska or Scarborough. Hopefully, Herb Adams' discovery in the *Portland Evening Express* would lead to a clear answer.

As soon as I got off the phone with Hal, I rushed to the microfilm section of the State Library, pulled out a box marked *Portland Press Herald*, Sept.-Dec. 1959, and fed the film into a reader. The screen flickered then stopped on a blurred image. September 1's front page came into focus. A large photograph appeared under the headline, "Prehistoric Find?" In front of the drooping arm of a large backhoe, two men in dusty work clothes knelt over a pile of sand. Between them lay bones, which looked like partial ribs, and in front a large tusk stretched across the sand.

Courtesy of the *Portland Press Herald.*

My pulse raced. The tusk looked like the Institute's specimen, but I couldn't be sure of its exact size and curvature, because the photo had been taken at three quarter view. Could it have been misidentified or was I looking at an entirely different tusk? I read the caption. The two men were James Littlejohn and Leonard Cash.

After printing a copy of the photo, I thumbed through the phone books. In the Greater Portland Directory I found the name Littlejohn, but not James. William J. Littlejohn seemed like a sure bet as a close relative so as soon as I got to my office, I called.

5. Thank You, Mr. Littlejohn

I slowed the museum pickup when we approached a mailbox. "That's Littlejohn," said my companion.

Bob Lewis worked as Bruce Bourque's archeology assistant at the Maine State Museum and had been in that position since before construction of the present Library – Museum - Archives complex in 1971. Many professionals regarded Bob as one of the finest field archaeologists in the state, and precisely because of his experience and critical powers of observation, I had asked him to accompany me on my trip to Scarborough. Besides, Bob's soft-spoken easygoing manner and great sense of humor made him the perfect traveling companion.

The week before, I had spoken by phone to William J. Littlejohn and his wife, Janice. Mr. Littlejohn was, in fact, the son of James who had died several years earlier. Fortunately for me, he played no sideline role in finding the tusk, and he displayed a vivid memory of that experience.

Mrs. Littlejohn cheerfully invited us into the house and directed us into the living room where we shook hands with Mr. Littlejohn, a slim, gray haired man with chiseled features. For Bob Lewis's sake, I asked Mr. Littlejohn to repeat the story he had told me.

When I first spoke with William Littlejohn by phone, he amazed me with the details he could recall about the tusk's discovery. However, to be objective, I needed to corroborate his story with others. Luckily, I located Leonard Cash and Christopher Packard. Both of them confirmed the parts they knew of Littlejohn's story. Furthermore, the 1959 *Portland Press Herald* articles that I had found in the state library and additional reports Herb Adams had sent me from the *Portland Evening Express* dovetailed with what I had heard.

One thing I focused on that all three men independently mentioned was damage to the tusk. The news photo showed no such damage. Had it been superficial enough to be overlooked? Did a master craftsman repair it for exhibition at the Portland Museum of Natural History? I got my answer from Christopher Packard when I asked, "The Portland collection contained two tusk specimens?"

"Yes," he replied, "one from Alaska, which was a monstrous thing, and the much smaller Scarborough specimen."

At about ten feet in length, Hal's tusk could hardly be called small. I felt disappointed but not surprised when Bill Littlejohn looked at my photos of it and said, "That's not the one we dug up."

Janice then handed me several color slides. "You can borrow these," she said, "but we'd like them back. We don't have many pictures of our daughter. She died very young."

I held a slide up to the light and saw an attractive teenage girl kneeling beside her younger brother. Spread before them on the grass lay ribs and the tusk, which looked about half the length of the Institute's specimen. Though the broken tip had been placed back on the body of the tusk, the damage was obvious. A wound of dazzling white glowed against the sepia tone of surface ivory and the smudges of gray clay. The size, damage, and sooty stains were all features that Hal mentioned when describing the tusk that still resided in Presque Isle.

Mr. Littlejohn drew a map of the site as he remembered it looking in 1959. The tusk, he felt certain, had been found near the north wall of the pit, leading him to believe that the bones of the beast still lay buried in the clay of that wall. When Mr. Littlejohn handed me the drawing, he invited us for a ride.

16

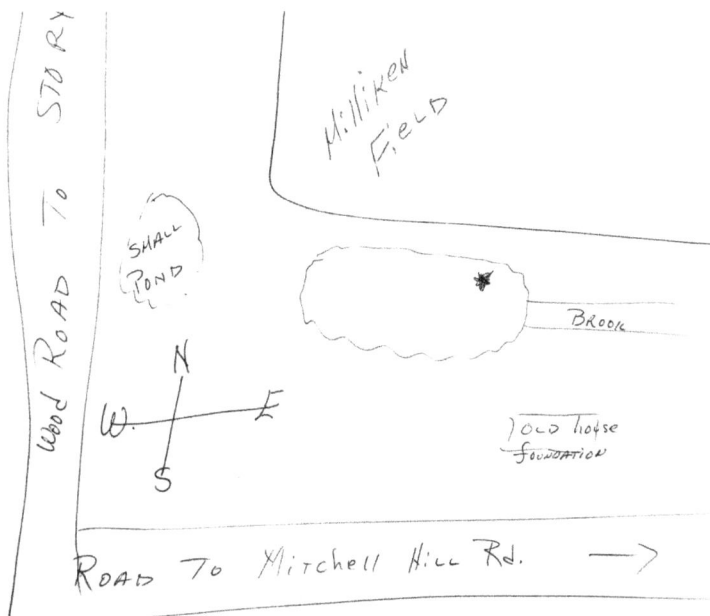

William Littlejohn's map of the pond location. The hand drawn asterisk represents the tusk's location in 1959.

In a short time Bob and I stared from the back seat of Littlejohn's car into a darkly wooded gulley choked with shrubs. We had parked half a mile from the highway on a well-maintained logging road.

Mr. Littlejohn pointed to a glint of sunlight in the forest. "That pool is where we took out fill for Ram Island. The pond you want is farther up the gully."

Armed with a note pad and camera, Bob and I waded through the underbrush, stopping briefly at the first little pond to watch a cloud of tadpoles race into the shadows. We climbed higher and broke into a clearing bathed in cool, luminous light. From the steeply pitched bank, ferns and brambles cascaded beneath us to the shore of a slate green pool thirty feet long.

While I shot a few frames of film, Bob scouted the surrounding area, checking on the lay of the land. Moments later he returned smiling. "There's a small stand of young Norway pine off to our right. It borders the wood road. Wouldn't take much cutting to give us good access to the site."

"Sounds good," I replied.

Bob chuckled. "Couldn't be better."

We returned to the Littlejohn home, and before leaving for the museum, Bob and I examined a box of seashells that had been collected near the tusk when it was discovered. Bob pointed to a tiny, oval shell. "*Portlandia arctica*," he exclaimed. "They're post-glacial. Today the southern limit of their range is Labrador."

I nodded, but inside I cheered. Post-glacial meant ancient, a time when the last mammoths and mastodons roamed North America.

6. "Extremely Unstable..."

On a warm summer day in 1990, I waited on the cement steps of a small wooden building at the University of Maine at Presque Isle [UMPI]. Geology Professor Kevin McCartney talked to me about his plans to organize a museum within the Science Department. His animated enthusiasm would bring this dream to fruition in a few short years. But for me the goal lay on the other side of the door where some of the Portland Society of Natural History Collection had been stored.

Brian Sipe, Maine State Museum's Curator of Decorative Arts, stood to one side of me. I asked him to make the trip with me because of his familiarity with the campus and key faculty members, having earned his undergraduate degree there. He also possessed the kind of chipper personality that won him many friends in and around Presque Isle, a city that he dearly loved.

Kevin unlocked the door and shoved it open. Boxes piled one on another lay over most of the floor space. In one corner the spiral tusk of a narwhal leaned against a cabinet. An alligator and leatherback sea turtle sprawled on the floor near us. But to one side of the turtle lay a dirty, broken object that riveted my attention. I instantly recognized the Scarborough tusk. The clay smudges, sienna surface color and chalk white damaged ivory that I had seen in the Littlejohn photos confirmed it, and beside it lay the broken tip.

Kevin and Brian became instrumental in working out a formal transfer of the tusk, but while that proceeded, Chief Conservator Stephen Brooke and Objects Conservator Madeleine Fang examined the tusk in their laboratory at the Maine State Museum to determine the object's physical condition and to formulate a stabilization treatment.

19

When the tusk came out of the clay in 1959, groundwater saturated the ivory, and though it looked pristine, its internal framework had partially degraded. The weakened structure only became apparent when the tusk broke. A newspaper at the time described the tusk's condition as "exceptionally well preserved with the marrow of the tusk still in a spongy condition."[1] Marrow is found in long bones, not tusks. The reporter had described the degraded condition of the ivory core. The tusk then lay in a garage for several months where wide temperature and humidity fluctuations persisted. After the Littlejohns donated it to the Portland Museum of Natural History, the tusk sat in a very dry atmosphere for many years. In effect it dried like a stick of firewood with a punkie center. At first the cross section of wood only appears darker than usual in the middle, but eventually a cluster of cracks develop. Thoroughly dry the log may look pristine on the surface, but tilting and a light tap will cause it to rain chips and wood dust. So when Steve and Madeleine examined the tusk, it presented a major conservation challenge as stated in their official report.

"Extremely unstable interior. The growth planes have cracking with major loss."[2]

Steve recommended that if we received title to the tusk it should be stored permanently in a stable environment. In other words, it should not be exposed to the fluctuations of temperature and humidity common in exhibitions.

7. Empirical Proof

A few weeks after Steve's assessment of the tusk's condition the museum had legal ownership of it. When Bruce Bourque offered to send ivory samples to a laboratory that he relied on for radiocarbon dating, I jumped at the offer. I gathered up some of the chips that spilled out of the tusk during transport from Presque Isle, and Bruce mailed them the next day to Geochron Laboratories Division of Kruger Enterprises, Inc. in Cambridge, Massachusetts.

Three forms (isotopes) of carbon exist in nature – C^{12}, C^{13}, and C^{14}. The numbers indicate atomic weights or how many protons and neutrons are in the nucleus of the carbon atom. C^{12} has the smallest number, 6 of each, which accounts for its atomic weight of 12, making it the lightest, and C^{14} with 8 neutrons and 6 protons being the heaviest. C^{12} and C^{13} remain stable, but C^{14} radioactively decays to a form of nitrogen gas. Plants take up these forms of carbon as carbon dioxide and incorporate them into their vegetative structure. When animals eat plants, they in turn integrate these forms of carbon into their body structures. At death their remains no longer take in carbon. So in our Scarborough tusk, C^{12} and C^{13} remained locked in the ivory, but C^{14} diminished over time as it continued to radioactively decay into nitrogen that dissipated into the atmosphere. Because of this predictable process, an accurate dating method had been devised for organic remains up to 50,000 years old – a point beyond which too little C^{14} remained for accurate dating.

I hoped the radiocarbon analysis would prove our elephant ancient and silence a rumor that surfaced shortly after the tusk's discovery, claiming it to be from a modern elephant. Some people who had been to the Scarborough site in 1959 remained convinced that the tusk was less than 150 years old, but they couldn't give me a good reason why. Even some of the geologists believed it. Bill Littlejohn's practical statement gave me hope. "Nobody would bury an elephant in clay."

While we waited for the dating results, Steve and Madeleine consulted by phone with ivory experts all over North America. Museum Director Paul Rivard and I felt strongly that if the tusk proved to be prehistoric, it should be kept in public view. With that in mind, the conservators investigated all possible methods of stabilization, which would allow the tusk to be safely displayed.

The goal of a conservator is to stabilize objects by using materials that prove totally reversible. When he or she replaces lost paint on a piece of artwork, the new paint is meant to be of a type that can be removed anytime in the future without damaging the original artwork. This is also true of glues, fillers, supports, and chemicals that contact the object being stabilized. Should future conservation work be required, the reversibility of previous restorative techniques becomes vitally important. However, this ideal cannot always be followed because totally reversible methods and materials do not yet exist for all problems encountered by conservators. This became the source of frustration for Steve and Madeleine in developing a stabilization method for the Scarborough tusk. After all the phone calls, they came to the conclusion that no reversible method existed.

As curator, I assumed the responsibility for the care of the tusk so I had to decide among the stabilization options Steve and Madeleine presented to me. I understood their reluctance to use a compromising technique, but if the tusk was to be exhibited, I felt that we had no choice. I asked the conservators to develop what they judged to be the least compromising technique that could be done within the six-month window we had before the exhibit hall opened.

On December 3rd conservators poured a clear polyester resin into the voids of the small tusk section, but later concerns about possible delamination of the resin over time made the conservators switch to a more reliable product.

On the morning of December 10th volunteer Judy Ritchie, under Madeleine's supervision, began infusing the large section of tusk with EPO-TEK 301-1. In theory this epoxy was reversible. In practice it is doubtful it could ever be leached from the ivory, but we judged it our best shot.

*   *   *   *   *   *

About a week later I noticed an opened envelope addressed to Bruce Bourque in my museum mail slot. In the upper left corner appeared the name Krueger Enterprise, Inc. I pulled out the analytical report: bone apatite – 10,125 ± 650 C-14 years BP (before present), bone gelatin – 10,935 ± 490 C-14 years BP. Both chemical fractions of the ivory proved the tusk ancient.

What a Christmas present! We not only had the only confirmed Ice Age elephant tusk from Maine, but as Hal Borns noted, "It is the only terrestrial vertebrate fossil known to have been found in Maine." [1] The tusk's stabilization would now take high priority.

Judy Ritchie had taken a keen interest in the tusk's consolidation.

Already the infusion process was well underway, but it required slow, delicate work. After having been proven ancient, the tusk consumed all of Judy's energy. By the time she completed her project, she had logged 120 hours, but she had compressed the work time into 12 days. That is a truly committed volunteer.

8. The Fisher of Ancient Elephants

After my completion of the archaeology hall's plant case, Director Paul Rivard gave me the rare honor of being responsible for the design, curatorial decisions, and construction of the introductory exhibit to the hall. Of the few Ice Age objects available to me, only two had the visual power to provoke the interest of visitors. They were pristine walrus skulls (male and female) found in Maine's coastal clay banks.

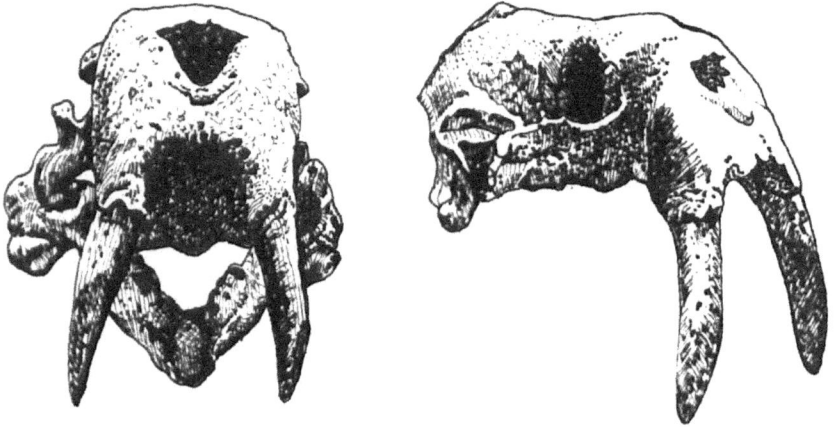

Walrus skulls. (Author's illustration)

Most of the other smaller objects had been dredged from the Gulf of Maine or had come from the old Portland Museum of Natural History, having no provenance. The only relic of an extinct species from Maine would be the Scarborough tusk.

Months earlier I had acquired a massive bone that would have been impressive in the introductory exhibit. Pulled up by a diver from the waters off Great Cranberry Isle, it was alleged to have come from a mammoth until Dr. Arthur Spiess, an archaeologist specializing in faunal remains, examined it. The bone proved to be the ulna from the flipper of a humpback whale, forcing a redesign of my exhibit space. [1]

After Judy Ritchie completed her ivory consolidation, Steve Brooke began building a padded support that perfectly conformed to the surface of the tusk. Now that we knew what objects would be exhibited, I concentrated on refining the fossil case.

Earlier in December Bruce had contacted an expert on ancient elephants, Dr. Daniel Fisher, from the Museum of Paleontology at the University of Michigan. At the beginning of his career Dr. Fisher worked as an invertebrate paleontologist, but he developed a strong interest in mastodons when he found that a common technique used to analyze the growth rings of clamshells could be modified to study ivory layers in ancient elephant tusks.

On December 27, 1990, the Fisher family arrived at the museum from Cape Cod where they vacationed with relatives. A casually dressed, bearded man wearing glasses introduced himself as Dan.

I had heard an old saying, "Whisper, if you want to get someone's attention." That expression could have been applied to Dan, except what enthralled me centered on the content of his soft speech while he scanned details of the tusk. He pointed to the cup depression at the tusk's origin – too shallow for a male. His hand slid along a band of ripples a quarter way down the main tusk section – signs of an infection or malnutrition. We propped up the tusk to his liking then held interconnecting puzzle pieces of ivory so Dan could estimate the length and degree of spiraling – unlikely a mastodon.

When Dan finished, he concluded that we possessed the 6 foot 8 inch left tusk of a female. And he felt 60% certain that the tusk belonged to a mammoth.

Dan then removed a small cross section of the wall of the tusk's pulp cavity so that he could determine the season of the animal's death. The last layer of ivory would have been laid down there, and Dan had correlated the ivory layers with the ratio of two different atomic weights of oxygen found in them. During summer when meltwater raged through the streams and rivers, these two forms of oxygen in water molecules mixed well, but in winter when water flow slackened, water molecules containing the heavier weight oxygen tended to settle into deeper water, causing drinking animals to acquire less of it. This effect became incorporated into the formation of ivory so that winter layers held a higher concentration of the lighter weight oxygen atoms than summer layers.

Before Dan left, he told us about his own research in the Great Lakes region, where he had found mastodon bones that appeared to have been butchered then stacked and weighed down with tusks and stones. His tusk analyses indicated the animals had died in late autumn. The soil profile led him to believe that the site had been a bog at the time of the butchering and meat had been cached in the acidic water for winter. To test that idea, he and his students butchered a recently deceased workhorse and stacked its meat underwater in a bog. They monitored the meat for harmful bacteria over the winter. By spring there was no sign of pathological growth, so Dan tasted it. He found that the meat had a "cheesy" but not objectionable flavor. The organic acids in bog water had inhibited decay lending significant support to Dan's meat cache hypothesis.

We had the tusk of an ancient elephant found in the clay layer of a tiny tributary with radiocarbon dates comparable to the time when paleo-Americans first entered the state. Was the Great Lakes scenario being repeated in Maine? Bruce expressed his eagerness to find out, but like me, and most of the museum staff, his commitment to finishing work on a major exhibit hall, scheduled to open in less than six months, remained top priority.

9. Yen and a bit of Yang

One day during the first week of 1991 while I worked on the introductory exhibit, Paul Rivard toured a group of dignitaries through the unfinished exhibit hall. He stopped and introduced me to a legislative lobbyist by the name of Kay Rand. She wanted to share some news. Her father, geologist Jack Rand, spent time at the Scarborough site in 1959 taking a home movie of the geologists working the dig. If the film could be found, we were free to make a copy for the museum. Finally, we might have "eyes" on the '59 find. She gave me her card, but before I called, a geologist from a renowned university gave me a ring.

A few days after receiving the radiocarbon dates on our ancient elephant, an Associated Press article about the discovery went to newspapers nationwide. At the same time, word of our find swept through the geological community. I expected Dr. ___ to know the news, but his call surprised me and so did his message. Dr. ___ had worked as one of the lead geologists on the 1959 Scarborough dig.

At last I was speaking with a professional scientist who might give me a clearer picture of what happened in 1959, particularly details of the depositional environment associated with the tusk, and what circumstances contributed to stopping the dig. Both William Littlejohn and Christopher Packard recalled the tusk being no deeper than a meter in the clay bed, and Leonard Cash described seeing a clay tubular half-mold of the tusk slumped in his bucket just before noticing an ivory glow in the pit. But I wanted the backup of hard data from a scientist.

Unfortunately, Dr. ___ provided little new information. When I mentioned Dan Fisher's assessment of the tusk probably being from a mammoth, he said that (Harvard paleontologist) Patterson had the same suspicion in 1959. I asked him why no one reported it. His answer paralleled those of everyone I had interviewed. A story circulated that the tusk might be from a modern elephant. He added that since researchers lacked funds for radiocarbon dating, the tusk's age could not be proven. I then mentioned that Jack Rand had a film of the '59 dig to which he replied that he knew Jack and he'd be glad to contact him for me. Already committed to a tight exhibit schedule, I thankfully accepted his offer, but before I hung up I told him Bruce would be interested in talking with him from an archaeological perspective.

*   *   *   *   *   *

By mid-January the 12,000 Years in Maine Hall was rapidly taking shape with the major exhibits nearly finished. But the majority of the smaller cases remained empty, instilling a great unease about our rapidly approaching May deadline. I lacked time for research other than periodic discussions with Bruce, Dan, and Hal about the Scarborough site. Then on the morning of January 14, I received a letter from Jack Rand.

Dear Gary:

[Dr. ___] has called me to ask if I had a map showing the location of the 1959 mastodon(?) tusk discovery in Scarboro, and my search...has drawn a total blank.

He did say that you knew just where the site is, however, and I would appreciate it if you could mark the location on the enclosed [map].

He also said that he had told you about the movie that I had of the digging party...you are welcome to copy any or all of it...I'll thank you kindly for the map location.

Jack Rand.

Great news, but I thought it odd that Dr. ___ hadn't asked me about the site's location. Months earlier I had contacted the site's landowners, and they made it abundantly clear that we must not reveal its location to the public. I consulted with my colleagues but could not mail my response until February 26th.

Dear Jack:

...Yes, we would be most interested in seeing your film and copying it to videotape. We are trying to locate additional bones associated with the site and are hoping your film will give us some clues...The dot that I put on your map is only an approximation, but it's certainly well within the neighborhood. The new landowners are concerned about their privacy, so please don't reveal the site's location to the general public or press...

By that time Bruce had been in communication with Dr. ___. He suggested that it might be wise to invite Dr. ___ to reinvestigate the site with our team. I liked the idea of a scientist with a history of the site working with us. I remained swamped with exhibit work and felt delighted when Bruce offered to take the lead in communicating with Dr. ___.

* * * * * *

We made our May 1991 deadline for the opening of the 12,000 Years In Maine Exhibit Hall. It proved a rousing success. The principal speaker for the event was an archaeologist from the Smithsonian Institution. Many dignitaries including representatives from the various indigenous tribes in Maine and guests from a number of other states and countries attended. For me, the highlight became a request by a British publisher to write an article on the creative work I'd done on the Native American Edible Plants Exhibit for his international journal published in Oxford, England.

The Native American Edible Plants Exhibit at the Maine State Museum.
Photo by Gary Hoyle.

But soon after the excitement, everything quieted down.

Following the frantic pace to meet our deadline, the lull seemed like a blessing at first, but in June a new state budget passed by the legislature hit the museum hard. Several key positions vanished, including Chief Designer, Chief Conservator, and Assistant Director. For the first time the museum charged visitors in order to stay open. Some friends I had worked with for nearly twenty years were gone.

The first part of July felt like a mourning period, but as time dragged on, the museum staff knew something else was up. In late July we got our answer. Our director, Paul Rivard, was leaving to become the Director of the Museum of American Textile History in North Andover, Massachusetts, a position he took that August.

10. A Bump in the Road

Shortly after Paul Rivard resigned, our overseeing board, the Maine State Museum Commission, appointed from its ranks Terrance Geaghan as Acting Director. Terry had some experience with museum collections and exhibit work, but to take the helm meant compressing years of knowledge and critical decision making into a nugget of time that required no interruptions from curators with pet projects like exhuming an ancient elephant.

I took advantage of the slow work pace in August to research 1959 newspaper articles about the tusk's discovery. From what I could tell no geologist had examined the tusk *in situ*. Only Christopher Packard was present when Littlejohn and Cash removed it, and the three men's stories complemented each other. Newspaper accounts paralleled their remembrances. The last 1959 article that I found about the discovery was written by Waldo Pray questioning the burial site of an elephant known as Old Bet that had been shot in Alfred, Maine, in 1816. Could the body of that elephant have been brought to Portland for display before being dumped into the Scarborough gulley? Pray posed a good question for its time, and that's what had stopped the dig. Even Packard had a similar concern early on. He eventually found that a circus had performed during early August 1816 in Portland with no mention of an elephant, but by then he was convinced the tusk was prehistoric.

*   *   *   *   *   *

One morning toward the end of August, Bruce and I sat in a cramped office filled with audiovisual equipment and shelves of videotape. Bill Dowling, audiovisual specialist for the Maine State Library, invited us down to view the tape he had just transferred from 8-millimeter film. Jack Rand's movie footage of the Scarborough site was ready to be viewed at last.

The segment opened on a murky pool where someone's head, fitted with a diving mask, bobbed on the surface. The sequence flashed to people standing around talking then slipping through mud, some of them driving probes and shovels into the slippery muck. The episode closed with a freeze frame of the tusk and ribs laid out on James Littlejohn's lawn.

In 3 minutes and 20 seconds, a visual history of what we knew raced by us. Over and over we watched, hoping to glean fresh insights. The overall impression of an unstructured excavation seemed apparent, but was that colored by a late twentieth century prejudice?

Bruce paused the tape on a large flat stone being lifted from the mud. Could that be significant, considering what we knew about Dan Fisher's research in the Great Lakes region? A probable meat cache had recently been found at Aziscohos Lake near Maine's border with northern New Hampshire and had been reassembled as a central exhibit for "12,000 Years in Maine." Was a cache of ancient elephant bones lying in the Scarborough mud?

Another idea seemed worth wrestling with. Had the elephant mired in mud? Had the site been a prehistoric watering hole for wildlife? A record of both modern and ancient elephants showed that entrapment in mud contributed to some elephant deaths. Dr. ___ championed this idea.

*   *   *   *   *   *

With our institution's survival threatened, the professional staff scaled back all projects. The priority turned into something none of us were versed in: finding a new director. Terry Geaghan guided us through the process, from formulating our vision of leadership, to reviewing resumes and finally interviewing candidates. The chance of excavating at the Scarborough site seemed as remote as the moon. But Terry had a novel idea.

Some of the projects Geaghan worked on in the past received assistance in engineering and construction from the Maine Army National Guard. If the search for our ancient elephant met with all the necessary criteria, we might be able to dig next summer in Scarborough.

In March of 1992 our new fulltime Director Joseph R. (JR) Phillips began his duties. Where Paul Rivard focused on the industrial history of Maine, JR embraced a more general vision, giving me hope he would support our plan to dig. Terry had already talked with Col. Joseph E. Tinkerman III, Director of Operations for the Maine Army National Guard. It looked as though our excavation would fit the guidelines of the 133rd Engineer Battalion, but a formal letter of request would be necessary. I now maintained more frequent communication with Hal Borns, Dan Fisher, and Bruce Bourque to tentatively plan for a scientific investigation of the site.

Just as plans coalesced, Bruce gave me shocking news. He had gotten off the phone with Dr. ___. After multiple discussions with our team, Dr. ___ reversed his decision to join us but was determined to publish our findings in a paper he would give at the Geological Society of America. As one of the original investigators of the Scarborough site, he certainly had priority to publish, but he did not have permission to publish our radiocarbon data.

I could do little about Dr. ___'s plans, and after an emotional discussion about it with my colleagues, we decided not to react. Even when he arrived at the museum weeks later to photograph the tusk, I treated him with the same respect that I would any other visitor. I led him to the Ice Age fossil case where without a word he thrust an open field journal at me. On one page was drawn a simple soil profile with the designations of a tusk and rib.

"May I take a photocopy?" I asked, and he nodded.

By the time I returned to Dr. ___, he had finished his photography. I escorted him back to the museum lobby and wished him well, then went to my office to study the photocopies.

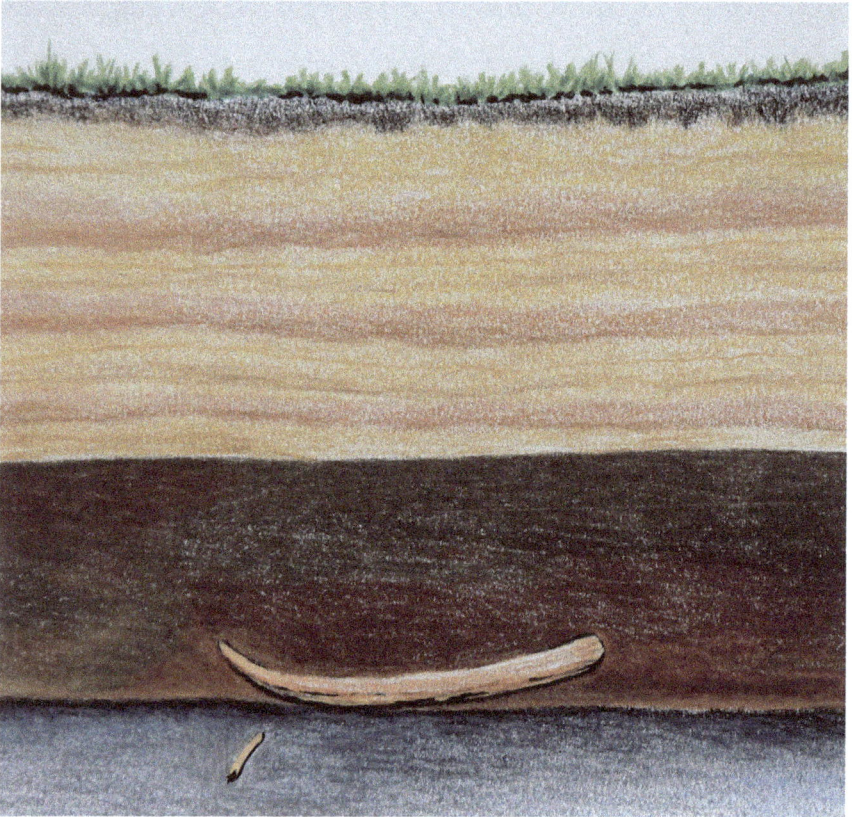

Author's rendering based on Dr.__'s field sketch. Note that Dr.__ located the tusk below three feet of bedded sands and within a two-foot depth of brown clay blanketing the gray marine clay layer.

Dr. ___'s journal contained little new information. In fact I was surprised at the limited amount of it and that his drawing located the tusk in what he called "brown clay" with its lower surface resting on the interface with "grey clay." Clearly, without our data Dr. ___ had little of significance to present at the Geological Society of America meeting.

11. Bundling Threads

In spring Terry Geaghan submitted a formal letter of request to the National Guard. JR Phillips also began negotiating with the Scarborough site landowners for permission to do an investigation of the area. They agreed to allow an excavation only if the reclaimed pond be built larger after the dig and press coverage not reveal the location of the site or the owners' names. We also needed to address any concerns of the Maine Department of Environmental Protection and the Army Corp of Engineers and to procure necessary permits.

Throughout the spring and into the summer JR Phillips and I worked with one of the landowners to develop an excavation plan that would benefit both parties. Our goals were compatible and straightforward: a complete scientific investigation of the site after which the pond would be enlarged to accommodate a population of stocked trout. But to achieve those goals within the 1992 field season required the orchestration of many factors.

Because the pond lay within a tributary of the Nonesuch River, no excavating could be done without a permit from the Maine Department of Environmental Protection [DEP]. The DEP also required us to hire a licensed pond designer, and in addition the Army Corp of Engineers needed to evaluate the site for its possible involvement. Normally this process required a few weeks, but the cuts in state funding stretched the time to months. In addition, the National Guard needed specific dates from us because its summer training schedule was filling up rapidly. If we couldn't determine an excavation date soon, we would have to wait another year.

The landowner who worked with us through these problems also sought permission from a closely abutting neighbor and the town of Scarborough to cut more trees at the site, so that we could have the flexibility of extending our search area to the property line.

While our new director and others worked out the initial legal and administrative problems, my effort focused on the scientific needs of the project. If fossil bones still lay buried at the site, they had a story to tell that required the trained eyes and experienced hands of geologists familiar with Maine's glacial history and Dan Fisher's knowledge of fossil elephant sites.

Scheduling the scientific people proved relatively easy. Dan Fisher agreed to be on call in Michigan when we found bone *in situ*. He would then fly to Portland, Maine, less than seven miles from the site. My geologist friend Woody Thompson and his colleague Tom Weddle of the Maine State Geological Survey agreed to visit the site on whatever the chosen date. Hal Borns had commitments during the first half of the summer. Bob Lewis, too, would be available only in late summer. His supervisor, Bruce Bourque, would be unable to attend no matter what the date. His field season was already booked solid.

After factoring in additional minor variables, it became clear that mid-August would be our earliest possible date to excavate, but by the time we made that decision the reservation time for the Guard had been booked. We rescheduled for September.

*   *   *   *   *   *

In early June I visited the site with archaeologist Bob Lewis, museum photographer-logistical man Greg Hart, and Jan Seleeby, husband of museum staffer June Seleeby. Jan was an experienced surveyor and offered to work with us at no cost. Over the course of several hours, Bob, Greg and I assisted him in establishing a datum point and additional points that described a precise, straight line across the pond's long axis. Bob marked the datum point with a stainless steel pin, all others we marked with wood. At the end of the day Jan's work provided us with a vital reference for accurately locating elements within the site. If we did find bone *in situ*, a gridded field could easily be extrapolated from Jan's line.

12. Gaining Experience

One mid-July evening in 1992 I stood at the top of a butte in the badlands of eastern Montana and looked down on a pocket of lights twinkling from the little town of Glendive five miles away. They were the only lights piercing the dusky violet land on all sides of me, clear to the horizon. A cooling breeze began, and I stepped closer to the cliff edge to catch the wind and avoid the sagebrush where rattlesnakes often hunted in the evening. A few nights earlier some of the women in our party had been startled by prairie rattlers striking out at them from bushes on the dirt road's edge. No one was hurt, but we quickly learned to walk the center hump during the rattlers' supper hour.

Author's field sketch of the 1992 tenting area in eastern Montana.

My two-week stint in this wild, eroding landscape was coming to a close. I had signed on to one of dinosaur hunter Jack Horner's projects as a volunteer to garner experience in bone recovery from fossil beds. Horner had become a well-respected researcher in his field, but he would soon be known more to the public as the science advisor for Steven Spielberg's Jurassic Park movies.

As a member of the survey team, I scouted for bone, walking along the eroding slope of a shallow canyon, while my teammates walked parallel above and below me. When one of us encountered bone, an alert went out for all of us to stop, while the team leader, Diane Gabriel, or her assistant, Karen Masta, extricated the bone and sometimes searched upslope for an eroding fossil bed. We worked mostly in a layer designated as the channel sands of an ancient shallow river. The bones from that area looked almost white and modern despite being 66 million year old remains of the last non-avian dinosaurs.

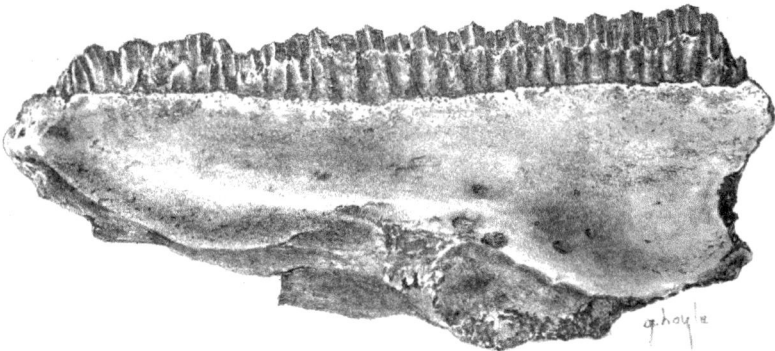

Author's field sketch of a duckbill dinosaur's jawbone.

Our finds would become a few data points in a research project spanning several years. The study would provide valuable information on the diversity of dinosaurs prior to their extinction at the end of the Cretaceous Period.[1]

For me the experience went beyond the academic. Thanks to Karen Masta, I learned the proper procedure for removing bone safely from the ground. I helped wrap several bones, even the rare, fragile remains of an immature triceratops. First, we partially exposed the fossils using trowels and brushes. A perimeter trench several inches beyond the bones was then dug in the soft rock matrix. Next, we covered the bones with a thin layer of plastic wrap over which we added a half-inch of wet toilet paper, followed by at least an inch of wet plaster bandages that extended to the edge of the trench. Digging then resumed until the rock had been so deeply undercut that the bones and their matrix rested on only a narrow pedestal. Plaster bandages were wrapped under the matrix up to the pedestal. A few taps of a chisel broke the pedestal so the entire mass could be rolled over and the plaster seal completed. By then the package could be safely transported to a laboratory for further work.

*   *   *   *   *   *

The laboratory responsible for the treatment of bones from the Scarborough site would be the Maine State Museum Conservation Laboratory where only technician Linda Carrell and volunteer Judy Ritchie now worked. No longer could I rely on the vast knowledge and skills of our conservators. Steve Brooke's and Madeleine Fang's positions had been eliminated months ago. Madeleine had taken a position at the University of California at Berkeley; Steve had become involved in a very different type of conservation, which would lead him to national fame. In 1999 his efforts contributed significantly to the removal of the Edwards Dam in Augusta, a major deterrent to the spawning sea-run fish in Maine's lower Kennebec River.

Linda's mild, soft-spoken manner belied the herculean task she faced, dealing with a vast spectrum of museum conservation issues, one of which was a plan for stabilizing whatever remains we found at the Scarborough site. She managed by consulting on the phone with conservators all over North America and then establishing a triage, so that objects for conservation treatment could be prioritized. Only objects with the greatest conservation need would be dealt with on our bare bones budget, and contracts for conservators would be severely restricted.

Fortunately for Linda and me, one of the country's finest objects conservators lived within fifty miles of the Maine State Museum. Ron Harvey had been the chief conservator at the Milwaukee Public Museum and now ran an independent conservation laboratory, Tuckerbrook Conservation, adjacent to his home in Lincolnville, Maine. His gregarious nature and experience in working on the complex challenges of preserving natural history specimens made him a perfect fit for our team. He would go on to work worldwide and distinguish himself as the caretaker of the famous fossil "Lucy."

Ron had worked on other projects for our museum in the past and became intrigued with our recovery of the tusk and its conservation challenge. He had been one of the conservators consulted for the tusk's consolidation. After the loss of our in-house conservators, Ron agreed to become an advisor on the project; Linda Carrell and her volunteer, Judy Ritchie, would be doing most of the physical work.

During this time Linda also began searching for additional information on ivory and bone preservation. She eventually recommended that I contact the Canadian Conservation Institute, an organization highly respected by conservators worldwide. Perhaps someone there would know about a recent wet site excavation of bones, similar to what we might encounter at Scarborough. Bruce Bourque had already asked several archaeologists and geologists about such sites and had come up empty-handed, so I doubted my efforts would be fruitful. But I was wrong.

Thanks to Linda's lead, within two days I was talking with geologist Robert Grantham and his assistant, Kelley Kosera, who worked at the Nova Scotia Museum of Natural History. They had just excavated the remains of an adult mastodon from a calcareous bog deposit and were developing a method to preserve the wet bones.

The method they developed to retrieve bone from their wet site roughly paralleled what I had learned in Montana. The thin plastic wrap over the *in situ* bone was the same, but moisture friendly spray urethane foam replaced the wet toilet paper cushion, further supported by a fiberglass resin mother-mold. The technique worked great for wet bone, not only as a support, but a method of limiting moisture loss.

Living bone exists in a high humidity environment within the body. When exposed to the air, drying creates distortion and stress fractures. If the drying is rapid, the bone can be completely destroyed because the extreme difference in humidity levels inside and outside the bone create enormous stress on its structure. The same is true for bone recovered from a wet site. The solution is slow drying. But how slow? In the early 1990s no one really knew.

Our Canadian friends decided to try a very slow approach. They hoped that cold storage would reduce the chance of mold growth as they incrementally ratcheted down the humidity. Unfortunately, mold did become a problem and the technique had to be modified to a more rapid drying method.

13. Launch Day

By September the wooden data stakes that Jan Seleeby plotted for us had weathered gray and the area around the little pond at the Scarborough site billowed with clumps of grass and sedge. The ninth of that month Bob Lewis and I waded through the green mass coated in morning dew, looking for the red marks that we had painted on top of each stake back in June. It began our first day of excavating the site, and I felt thankful Bob worked with me. I needed not just his expertise but also his calming influence. The months of planning and negotiating for the project's success weighed heavily on me.

By mid-day some of our team arrived along with newspaper, radio, and television reporters. Time then centered on our relationship with the media. To me it seemed vital that the people of Maine know exactly what we were doing because their taxes funded the project. I wanted them to experience our highs and lows. Every time reporters arrived at the site, I made sure we accommodated them. That first day Hal Borns, Lou McNally, and I gave our individual perspectives on the importance of the project. McNally, a meteorologist and well-known radio and television personality, came to the site as an advanced degree candidate under Hal. Though not part of our team, Lou gave a welcomed climatological perspective regarding our investigation.

The high point for Linda Carrell and me came a bit later when, to our surprise, the Canadian mastodon excavators, Bob Grantham and Kelley Kosera, arrived for a short visit before heading south to Boston for a major geological conference (the same conference where Dr. ___ would be announcing our radiocarbon dates on the tusk). We finally connected faces with phone calls and written correspondence. For months Bob and Kelley had been gracious about

45

sharing their field and laboratory experiences, so by the time we met in person, it seemed like I faced old friends. I wish I had been able to spend more time with them.

* * * * * *

As happy as I became about our favorable press coverage, I remained concerned about the possible disclosure of the site's location. The landowners demanded that their names and the exact whereabouts of the site not be made public. So I asked every reporter to honor the landowners' wishes. All except two complied. One threatened to release the information, to which I replied that his action would likely shut down the project. He never disclosed the landowners' names or the site's location. However, another reporter, probably forgetting his promise to me, described the exact whereabouts of our dig. For days I awaited repercussions over the incident, but they never came.

Linda Carrell and Judy Ritchie erected a small tent at the site and filled it with the supplies and equipment to establish a field conservation laboratory. Thanks to the advice of Dan Fisher, Bob Grantham, Kelley Kosera and Ron Harvey, we developed a reliable procedure for working with the wet bone that we hoped to encounter. A local food distributor even agreed to temporarily store any large elements of the skeleton in a cold room near the site.

* * * * * *

The pertinent reason for our day came in the late afternoon when Woody Thompson and Tom Weddle arrived from the Maine Geological Survey. Shortly after meeting with our National Guard crew and being briefed by Captain David Duehring, the geologists directed a backhoe operator to cut the first test pit fifty yards southwest of the pond where they examined the soil profile to establish the general history of the site prior to the 1959 disturbance.

The following morning the backhoe was moved to the north wall of the gulley near where the tusk had been found, and a new trench dug. Over and over Sargent Phil Caesar stretched the arm of the camouflage-painted machine

<Layered sand

< Brown clay

< Gray clay

Photo by the author.

down alongside the wall he carved. At the limit of its reach the claw peeled away a bucketful of blue-gray clay then rose without chafing the wall and dumped its contents at the feet of Hal, Woody, Tom and a small group of museum people.

"(The) geologists consider Phil an artist. His wall is as smooth as a plastered wall," I wrote in my journal.

We fingered through the pile of slippery clay, searching for a clue to its origin. Then someone called out, "Portlandia!" That shout confirmed that we were in ancient marine clay.

14. A Little Clam's Importance

About two and a half million years ago the polar regions of Earth cooled to the point where the accumulation of winter snows exceeded the summer melt. While the snow depth increased from years of buildup, ice crystals in the interior fused to form a single mass. It marked the birth of the Pleistocene Epoch, a time popularly known as the Ice Age though it was only the most recent of several ice ages in Earth's history.

As snow accumulated over the years to form ice, pressure became so great at the base of the load that ice there liquefied, and the entire load began to move. It had become a glacier that would slowly cover vast tracts of land and eventually merge with others of its kind. Throughout the Pleistocene these glaciers expanded and contracted numerous times.

The last Ice Age expansion began about 100,000 years ago. At that time a glacier developed in Canada's Hudson Bay that would become the dominant feature of the North American landscape for tens of thousands of years.

Just as cake batter spreads across a greased pan from its loading point, the continual accumulation of winter snows caused ice to flow out of Hudson Bay to all points on the horizon. Close to 25,000 years ago this glacier, which geologists call the Laurentide Ice Sheet, reached its maximum size, extending north into the Arctic Ocean, east to the continental shelf, and south as far as Missouri and Ohio. On its western flank it merged with a much smaller ice sheet, the Cordilleran, which flowed off the Canadian Rockies. With the exception of western Alaska virtually the entire area north of the present Missouri and Ohio River valleys had become an ice field. In parts of Canada the ice measured at least two miles thick.

The glacier profoundly affected the land. Its weight depressed the Earth's crust by several hundred feet, and as it crept forward, ice bulldozed soils down to bedrock in most places. Whenever any part of the glacier started down an incline, the reduction in pressure flash froze the lubricating sub-glacial water to the bedrock then plucked away portions of the ledge as the glacier moved on. These fragments ranged from slivers to house-size boulders. An example of the latter is Daggett Rock in the little town of Phillips near the New Hampshire border in northwestern Maine.

Daggett Rock. Author's illustration.

At an estimated weight of over 8,000 tons and a length of at least 80 feet, the Daggett Rock is the largest glacial erratic in Maine and one of the largest in the northeastern states. It's called an erratic because the transported stone is different from the surrounding bedrock. Geologists have determined that

Daggett Rock originated in the Saddleback Mountain Range, seven miles from its present location. Not long after settling, it split into three large sections.

When the glacier dragged rock fragments forward they ground against bedrock, smoothing the contours of ledges, scratching and polishing hard rock, and wearing trails in softer stone until these fragments ground to oblivion or were dumped miles from their origin in the floor of sub-glacial tributaries or at the glacier's front. In essence the glacier functioned like a colossal sanding machine using an inexhaustible supply of plucked rocks to shape the land.

Ice sheets are also intimately linked to the sea. The snow that accumulates to form glacial ice initially evaporated from the ocean surface. By the time the Laurentide Ice Sheet reached its maximum size, so much water became locked up in the glaciers worldwide that sea level stood close to 400 feet below the present shoreline. In Maine this situation became more complex than in noncoastal regions. The weight of the glacier had depressed much of the state's landmass well below the present shoreline.

At the time of glacial maximum, ice lay nearly one and a half miles thick over parts of Maine. The front of the ice sheet hung over a hundred and eighty miles east of the present Maine coastline at the location of Georges Bank in the Atlantic Ocean, where it calved icebergs into deep water. Over seventy thousand years had passed since it spilled out of Hudson Bay, but in a geologic instant it melted away.

The warming trend began about 21,000 years ago.[1] Though ice continued moving into Maine from the northwest, rapid melting eroded the ice front so that by 15,000 years ago the edge of the Laurentide Ice Sheet stood near the present Maine coastline. Torrents of meltwater dumped the tailings from thousands of years of rock grinding directly into the newly formed Gulf of Maine where the finest particles drifted for miles until chemically altered by seawater. These particles then clumped together and settled, to eventually form extensive beds of marine clay.

A short-faced bear encounters a female woolly mammoth during glacial recession. (Author's illustration.)

Over the next thousand years ice receded from the southern half of Maine. However, most of the meltwater still flushed directly into the sea because the landmass that once buckled to below sea level under the ice load now lay underwater. The land's rebound would take much longer than the glacier's recession.

As soon as the eroding glacier exposed new ocean bottom, life came to colonize. Only a few species could tolerate the floods of freshwater and heavy sediment load close to the ice sheet. One of those tolerant species was what geologists pulled from the clay in 1992 and what Bill Littlejohn had collected from around the tusk in 1959 – the little clam, *Portlandia arctica*.

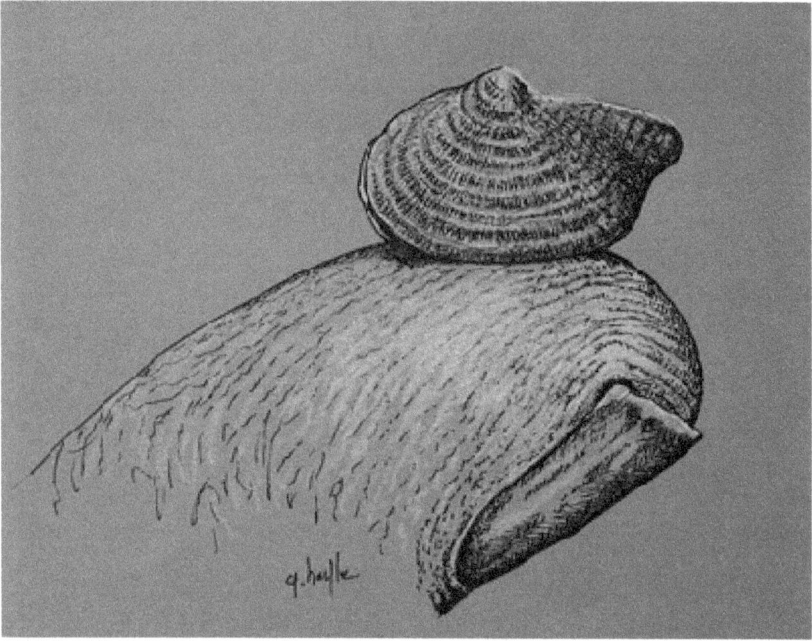

Portlandia arctica (Author's illustration.)

15. "Bone! We Found Bone!"

September 10, 1992:

"Geologists find layers of different textured sands of 2-2 ½ ft. depth over 2 ½ ft. of gray-green clay. 8 – 12" sharp gravel found below this clay before encountering blue-gray clay (classic marine). Water is pouring from gravel layer…Hal asks if we can continue cut all across north wall to end of pond."

The sun casted long evening shadows when I wrote those words in my journal. We had encountered the basic stratigraphy described by Littlejohn, Cash, and Dr. ___, but we needed a larger exposure to understand how the bones had gotten into marine clay. Caesar's cut into the pond's north shoreline was close to where the tusk had been found, and we felt reluctant to continue enlarging it without some type of safeguard to prevent the backhoe from damaging buried bones. That safeguard arrived in mid-afternoon.

Early in my discussions with Captain Duehring, I expressed my concern about the danger of blindly digging for fossils with heavy equipment. How could we reduce the risk of damage? Captain Duehring suggested monitoring our progress with a nuclear densitometer.

Basically a nuclear densitometer is a device made up of a shielded chamber containing radioactive material coupled to a sensitive metering instrument. When placed on the ground's surface a window is opened to allow radiation to escape into the soil. The metering instrument then reads the radiation's backscatter. After the operator calibrates the instrument by measuring at a number of known density points, the densitometer is ready for use. The operator then moves his equipment over an area in question and takes a reading at every placement of the densitometer until a grid pattern of readings produces a graph of variations in density below the surface.

The nuclear densitometer had been used for many years to check for subsurface defects on aviation runways, but it had also proven its worth for other purposes.

The guardsmen told of one incident where a woman mysteriously disappeared from her home. Her husband claimed that she had gone to visit relatives, but neighbors and the police were suspicious that something sinister had happened. For months the police searched for the missing woman with no success. Then they sought the assistance of the National Guard in a last ditch attempt to examine the couples' cellar. A nuclear densitometer positioned over a new patch of concrete picked up variations that, when plotted, took on the appearance of bones. Within hours the police dug up the remains of the missing woman and arrested her husband for murder.

By late afternoon Sgt. Joe Brocato had calibrated the nuclear densitometer in marine clay at the Scarborough site. We laid out our work schedule for the next day. Phil Caesar would continue his cut along the gulley's north wall while Joe Brocato monitored the clay exposed by the backhoe. I would photograph the process for Hal since he would be teaching that day and would be in Washington D. C. the following week. But before additional excavating could be done, a huge pile of earth dug up in 1959 would have to be moved. It lay close enough to the crest of the gulley to seriously compromise the safety of Caesar's excavation. Phil agreed to move it away from the site first thing in the morning.

*   *   *   *   *   *

Friday September 11 dawned clear, a beautiful day to excavate. Knowing that the small mountain of spoils would have to be moved before excavation, I took advantage of the time to drop by the museum before heading to Scarborough. I wanted to check my mail and update the director on our plans and progress. Fortunately, JR arrived early and the briefing with him lasted only a few minutes. Even at that I did not get on site until after 9:00 AM.

When I arrived, a haze of dust hung like fog. A fresh pile of earth lay dumped far up on the pond's south bank. That was where we intended to screen for clues about our fossil elephant much later. But plans suddenly changed.

A chorus of shouts met me as I jumped out of the museum truck.

"Bone! We found bone!"

16. Hairy-it

In eastern Montana I had been struck by the preservation quality of dinosaur bones in the channel sands of the upper Cretaceous. Some of them looked as fresh as the bleached cattle bones that we frequently encountered. When I discovered a dazzling white rib protruding from the wall of a butte, only the bone's size convinced me that it belonged to an ancient monster. It felt much heavier than modern bone due to mineralization during its 66-million-year entombment. But its color and surface texture looked, to my untrained eye, identical to a cattle rib that I'd found a few days earlier. If the bones in Scarborough looked this good, I'd be ecstatic. I never dreamed they'd look better.

Linda Carrell and Judy Ritchie led me on a brisk walk to the conservation tent where they opened a portable cooler. From a plastic bag one of them handed me what looked like a long, fat, earth-covered candy bar. When I rubbed away dirt, I felt the hard, waxy surface of bone. It was so unlike the bone of dinosaurs or even cattle that I'd found in the badlands. This felt alive – fresh from the animal. No wonder some people in 1959 thought the remains were modern.

My faithful elderly volunteer, Jim Story, had found the first fossil fragment, a rib section, right after the Guard truck hauled off the first load from the north bank. Jim's discovery stunned me. After months of negotiation and planning, one of the fruits of our labor lay in my hand in such a short time at the site. But the cooler held other treasures. Our nuke man, Joe Brocato, had picked up something that made me shiver when I lifted it from the cooler. The double-fist size mass of dirt and bone sprouted root tendrils on one side, but on the opposite face a quarter-moon cup of bone curved toward me, and a swollen arch protruded from the dirt. It was a fragment of the base of the skull where the backbone connects. Known as the occipital condyle, I hoped it might prove to be a diagnostic feature.

Phil Caesar offered me the use of his car phone, and as I rushed to call the museum, Bob Lewis pulled out the skull piece and began flipping through a bone picture atlas of mastodon and mammoth skeletons.

I reached JR in his office. I wanted him to hear the news directly from me and not just because he was the director. He and I had worked hard over the many facets of the project. He too felt surprised and excited that we had found bone so quickly and anticipated that we would soon identify our mystery elephant.

I then called my wife, Jeanne. That's when I realized how emotionally connected I had become to the project. The six-year-old kid in me surfaced, exuberant about the discovery, gushing details of the exquisite bone preservation like a hack poet.

Though we couldn't be certain of our elephant's identity, the scientific evidence favored one species. Bob turned to a page in the atlas that compared skull bases. The arch of bone conformed to a mammoth's occipital condyle. Was it a diagnostic feature? Some variability exists in all animal populations. I knew only one skeletal element that would solve our puzzle. As Christopher Packard stated in 1959, "...the molar is the thing." [1]

Before our bone discovery I assumed the elephant would prove to be a mastodon. Considerable scientific evidence supported the view that as the Laurentide Ice Sheet receded from Maine and land began to rebound above sea level, forests rapidly invaded, leaving little or no grasslands available to support a population of mammoths.

Though our evidence seemed tenuous at the site, it supported Dan Fisher's earlier assessment of the tusk. Perhaps Maine supported grasslands for a short window of time, but September 11, 1992, offered no time to speculate. More bone needed to be found.

On the north bank I picked up what looked like the broken end of a tree root and felt the elation of discovery as I lifted a rib fragment in the air. Then Phil Caesar found a rib fragment and became so excited that he wanted to lose his tickets to Panama and stay on the dig. But military orders ruled, and next morning the geologists' "artist" went on his way out of the country.

Linda, too, found bone – a section of the skull. Then Judy offered a name for our beast. What about Harriet? Dan Fisher described the tusk as characteristic of a female, but we were not yet certain she was a mammoth. I thought a moment. What about Hairy-it? The name stuck.

That night when I turned on television news, I watched WCSH-TV's Lee Nelson interview me. I saw activity at the site, smiling people with bone fragments, Bob Lewis holding the occipital condyle against the key page of the bone atlas. It was great to see, but it all flashed by too quickly. Then I thought about a promise Judy made that day. She'd kiss whoever found a tooth of Hairy-it. I chuckled to myself. It had been a good day, a very good day.

17. Nailed It Then Got Nailed

Though the guardsmen did not work the following day, which was a Saturday, Jeanne and I went to the site to sift through dirt with several volunteers. Everyone found bone fragments including one of the property owners who had taken an interest in the dig. It gratified me to see people experience the thrill of scientific discovery, but one person stands out in my memory of that day. He was a college student named Brian, and when he found a section of skull the size of two grapefruits, he became so overwhelmed that I noted in my journal, "His face goes dead white!"

On Monday commotion erupted again at the site. Judy met me as I got out of the truck but refused to tell me anything. She led me to the conservation tent and popped open the cooler. There lay the thing Christopher Packard had tried so hard to find, the thing I had hoped and dreamed about: a single tooth double the size of my fist, weighing 7 ½ pounds. The washboard appearance of its biting surface provided the confirmation I needed. We had indeed discovered a mammoth.

The following Friday's Portland Press Herald credited the tooth's discoverer. Jim Story came through again and got his kiss from Judy.

Within hours television, radio, and newspaper reporters swarmed the site eager to record the significance of our discovery. We had found the only mammoth known to science from Maine. The discovery generated a wave of enthusiasm that swept through the general public and scientific circles. The Maine State Museum became inundated with phone calls from people of all

The author holds a tooth of Hairy-it. Its discoverer, Jim Story, stands in the background. (Courtesy of the *Portland Press Herald*.)

walks of life wanting to know more about the mammoth and how they could get involved with our project. An army of new volunteers arrived at the site. Nearly overnight we had been thrust into the limelight, a blessing that soon proved to have its downside.

The ensuing days brought more bone and ivory fragments to light though none as spectacular as the earlier finds. Nevertheless, we knew the importance of every piece. Each was not just a puzzle chip of Hairy-it but a potential clue to the prehistoric past. A stain, puncture, or abrasion on any fragment might tell volumes to the trained eye. So I became particularly concerned one morning when an eight-year old boy, who had been at the site before any of our crew arrived, showed me his souvenir – a sliver of Hairy-it's skull. He gave me the fragment, but it reminded me how vulnerable the site had become to pilfering. Too many people knew our location, and each night the site remained unsecure. Furthermore, the guardsmen would leave us in a few more weeks. Until that time they would concentrate on stabilizing the pond's shoreline for winter.

Part of the stabilization process authorized by the Maine Department of Environmental Protection involved planting rapidly growing grass over the entire site. To accomplish it within our dwindling field season, JR contacted the Maine Department of Transportation [DOT]. The DOT had several pieces of heavy equipment specially designed to spray a mixture of water, fertilizer, grass seed, and a green biodegradable adhesive. The process, known as hydroseeding, had become the standard for stabilizing road banks.

But how could we remove all the skeletal fragments from the piles of earth before the ground froze? For us to abandon the site until the following spring would be like an open invitation to souvenir hunters. My logistic man, museum photographer Greg Hart, developed an ingenious plan to solve that problem. From a local contractor he borrowed a four-foot square, heavy-duty screen used to separate stones from fine gravel. With a mesh size of one inch, the screen could rapidly sieve our mixture of clay and sand without losing significant pieces of bone and ivory. Greg conceived the idea of mounting it at a 45-degree angle in a wooden frame directly under a hopper. The hopper would receive earth from a small front-end loader then funnel it onto the screen where spotters would carefully comb dirt down the screen and pick out fossil bits. While he and museum carpenter Scott Mosher built the apparatus, trucks laden with fossil earth were headed toward Augusta.

JR had made arrangements with the DOT to move the Scarborough spoils to an undisclosed location – a highway maintenance lot in the town of Manchester, less than five miles from the Maine State Museum. By Thursday, October 24th, the last of ten truckloads of our precious dirt pulled away from the mammoth site. At last the fossils were hidden from the public eye, and I breathed a sigh of relief, but that was short lived.

Before the day was through, one of the landowners appeared at the site. He thrust a paper into my hand.

"What's this?" he asked.

A courtesy museum newsletter, asking support for the mammoth excavation, had arrived. A photo of the landowner and me overlooking the pond prior to our investigation dominated the cover page. That was the moment I learned the museum's understanding of "public" was not the same as the landowners'.

18. Plotting a New Attack

A light wind sifted snow across the road on the morning of February 12, 1993, when I approached the turnoff point to the Scarborough site. I parked in front of the entrance gate and kept the truck running. The outside temperature stood at -5 F ^{0}. Ahead of me sunlight raked through the woods leaving the logging road striped in blue and shocking white. The road had been recently plowed for woodcutters, and trees screened the wind enough to prevent snow from drifting back onto the gravel.

I looked at my watch – 8:10. I arrived first at the rendezvous point, but I knew that Greg Hart wasn't far behind me. He'd be coming in his own vehicle so that he could head south by early afternoon to visit his daughter at an out of state college. As for the two people coming from New Hampshire, I hoped they could keep to their scheduled time and our directions would prove adequate for getting them to the site.

*   *   *   *   *   *

Much had happened since we relocated the piles of Scarborough spoils to Manchester. As soon as the earth arrived in late October, Greg Hart and Museum Shop Foreman Scott Mosher set up the processing screen and, together with JR Philips, made arrangements through the DOT for use of a small bucket loader and operator. Throughout the following weeks, museum staff members and volunteers raked dirt through the screen and picked off bone and ivory fragments, while Greg, Scott, or the DOT technician fed the hopper.

The process proved efficient. By the end of fall we had screened all one hundred cubic yards of earth and recovered nearly 300 skeletal pieces including another of Hairy-it's massive teeth.

All the fragments remained coated with clay and sand, requiring thorough cleaning with utmost caution. Scraping with metal tools risked marring the bone's surface or possibly obliterating marks left by teeth of an ancient killer or scavenger. Softwood probes and ox-hair brushes became the preferred tools for the job. With them Linda Carrell's team of museum staff and volunteers logged several hundred hours in the museum's conservation laboratory, gently removing all traces of dirt from bone and ivory.

Dirt removal was not the only challenge for the conservation team. Soil moisture kept Hairy-it's remains at a relatively stable, high humidity. But once removed from the ground, those remains entered a far more hostile environment, where rapid moisture loss became our primary concern. The tusk proved to all of us what rapid drying could do, as we watched it rain ivory when it arrived at the museum. Without Judy Ritchie's resin impregnation, the tusk would have crumbled to bits over the years. We vowed not to let our excavated specimens suffer a similar fate.

As soon as we found a portion of Hairy-it, it went into a plastic bag with soil. The bag then went into a cooler for transport to the museum. At the conservation laboratory, skeletal elements were cleaned then transferred to glass covered trays containing silica gel. Ordinarily silica gel is used as a desiccant, but Linda and Judy found it effective in controlling humidity levels, thanks to research that had been done by Ron Harvey. By adding specific amounts of water to the silica gel, Linda and Judy achieved fairly stable bands of relative humidity from very low to very high levels. They chose to create a relative humidity of 45% in the trays. It would take many months for Hairy-it's remains to dry to that level, but the stress on them would be minimal. The alternative was to store bones and ivory at a higher humidity, but that risked attacks by mold, which had become a problem for our friends in Canada with their mastodon specimen.

Having the cleaned bones under glass allowed us to get an overview of our collection. Many of the fragments were small – two inches or less, but there were impressive pieces of the skull, ribs, and two enormous teeth.

I felt elated. In less than four months we accomplished an amazing amount of work. Not only had we proven Hairy-it to be a mammoth, but we had recovered an abundance of fragments hinting that more of the animal might remain buried at the site. Only a systematic excavation could provide the answer. Bruce Bourque thought we could greatly increase our probability of finding bone in the clay bed by surveying with ground penetrating radar. He had already contacted experts in that technology at Geophysical Survey Systems, Inc. [GSSI] in North Salem, New Hampshire.

*   *   *   *   *   *

Shortly after Greg parked behind me on that cold February morning in 1993, Dan Delea and Jerry Pageau arrived from GSSI, and we headed to the mammoth site.

The pond lay in shadow when we arrived. Subsurface seepage and late fall rains had replenished it to its original surface level. The clearing above it dazzled with light. Because of the previous day's warmth only an inch of snow covered the ground, enabling the GSSI crew to park their vehicle at the pond's edge. When Dan Delea got out, he looked at the ice and pronounced it the best gridded field he had seen.

Weeks before the GSSI team arrived, we began planning for the radar survey. It was obvious that in order to have an accurate plot of the pond, a gridded field was needed. To accomplish that I envisioned using a scaled-down version of a device used by commercial ice cutters in the 19th century.

The ice marker was a cutting tool used mostly on rivers to layout the pattern of an ice field. It consisted of a single row of cutting teeth, which plowed a few inches into the ice, parallel to a previous line cut by a smaller, hand-driven marker. A gauge sliding through the channel of the first cut insured the precise alignment for the marker. When a horse pulled the tool along the ice, an operator held the unit steady by way of widely spaced handles. When the gauge reached the end of the first groove, the marker was turned around and the gauge flipped over into the second groove. A third groove was then cut and subsequently a series of parallel lines marked the field. Next, the operator, in the same way, scored cross lines until he had formed a grid pattern that established the size and number of ice blocks that would be removed from the river.

For our purpose I needed an ice marker small enough to be easily transported but heavy enough to cut efficiently when pulled by a person. I also needed the gauge set at exactly a half meter. With that in mind, I met with museum carpenter Burt Truman. Burt was a superb craftsman and a wizard at solving the kind of technical problem I presented to him. In less than a week he had created the tool we needed.

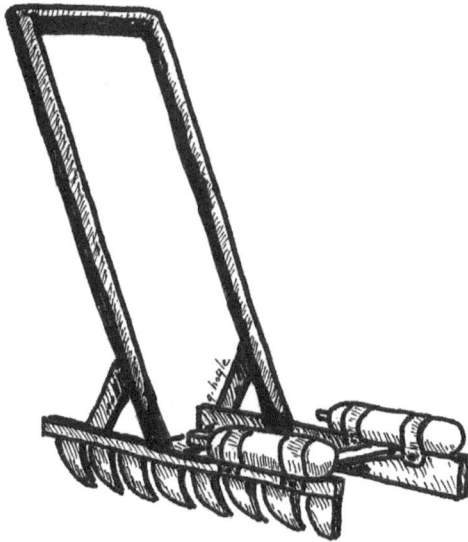

Burt Truman's ice scriber. Note the antique window sash weights. (Author's illustration.)

The day before the radar crew arrived, Bob Lewis, Greg Hart and I took Burt's creation to the mammoth site. For much of the winter the ground had been snow free, but a recent storm had dumped a few inches of dry powder that still covered the land. When we rumbled down the logging road to the site, we were prepared for a long day of snow shoveling and ice scribing. But the pond's surface lay snow free.

My most faithful volunteers at the site, Jim and Evangeline Story, lived close by. Nearly every day of our fieldwork they arrived at the site sooner and left later than the rest of us despite their advanced ages. Their dedication brought to light many significant pieces of bone and ivory and the first tooth.

Occasionally I stopped and visited them in their home, where I was treated to homemade pies and other goodies. I often kept the Storys informed about many aspects of the project, so the day before I left for the site, I telephoned to let them know about our plans. To my surprise Jim had shoveled the snow from the pond before we arrived.

In little time we located the brightly painted stakes that we had driven into the ground the previous summer. From them Bob stretched a taut rope across the ice to establish our baseline. Guided by the rope I slowly pushed Burt's ice marker over the frozen surface. Again and again I went over the line to ensure we had enough depth to seat the gauge. Once satisfied with the baseline, Bob and I took turns scribing parallel lines while Greg shot videotape of the process. At times one of us tried pulling the marker with a rope as the other guided it, but we lost control and soon resorted to pushing the device.

When we began marking, temperatures hovered in the low teens and the hard, brittle ice resisted our work, but as the day warmed, Burt's marker glided across the pond, plowing up hills of frosty shavings. To insure that the lines remained visible for the radar crew, we sprinkled red chalk dust into the grooves then swept the ice.

The following morning while Dan Delea and Jerry Pageau admired our work, Greg set up the video camera and within an hour he was shooting the slow, steady movement of a strange sled pulled across the ice.

Two sleds were used consecutively, both containing devices to send and receive radar through a "dish" directed at the ground. One sled sent a signal at 300 megahertz, the other at 500 megahertz. The 500-megahertz unit was preferred for penetrating most soil types because its higher frequency could penetrate deeper. But there was good reason to use both sleds.

Not all soils are good candidates for ground penetrating radar; one of the worst is clay. Traditionally, when farmers constructed an electric fence to restrict animals, they used porcelain insulators to insure the charge wouldn't be lost to the ground through the fence post. The clay of the insulator blocked the electric current. A natural clay bed acts in a similar though less efficient way due to its high moisture content.

When the radar scanning began, Jerry Pageau slid the 500-megahertz unit over the ice, as Bob Lewis and I watched the graph recorder with Dan Delea. While Jerry moved along the ice, we searched the screen for anomalies in the pond bottom. A scatter of colored squares rolled by, and the recorder committed it to monochrome on paper.

"It doesn't like the clay," Dan said. "The signal is being deflected."

An hour later Dan removed the chart recording and placed it in his briefcase before swapping places with Jerry. This time the 300-megahertz sled glided across the ice, and again colored squares rolled across the monitor. But now there was less reflection, allowing us to see deeper: only 2' – 2 ½' into the pond bottom, Jerry estimated. Both Dan and Jerry became disappointed that we could get no deeper, but I thought it might be enough to detect bone.

As Dan finished his course with the sled, Jerry pointed to areas on the chart paper and said they might be significant anomalies. Only a specialist in the North Salem office could determine it for sure, and it might take a month before we received his report. For the present I was just happy the pond bottom got surveyed, and we could get out of the bitter cold.

19. Clearing Muddy Water

On Friday, March 19, 1993, a momentous event took place at the Maine State Museum. Known as the "Mammoth Workshop," it originated as the brainchild of Bruce Bourque.

Bruce came up with the idea during one of our Hairy-it project meetings. His years of experience networking with other scientists led him to the conclusion that, with the exception of Dan Fisher, all the talent needed for the success of our project could be found in Maine. That sounded great to me. My own experiences with Maine researchers had been positive any time I needed advice on collections or exhibit work. To invite them to a workshop at the museum for advice on aspects of our project seemed a wonderful way to utilize their intellect and honor their contributions to the museum.

That morning while Linda Carrell set up a display case of the larger elements of Hairy-it's skeleton, JR and I met with Hal Borns to get an understanding about publishing and protocol. After the meeting I noted in my journal, "I certainly am getting an education in ways that I never considered."

At 11:00 AM we had a team meeting in the museum's conference room. The principle investigators had already been chosen: Bruce Bourque - archaeology, Hal Borns – geology, Dan Fisher – probosideans (elephants). Others at the meeting besides JR and me were Woody Thompson and Tom Weddle of the Maine Geological Survey and Bob Nelson from Colby College. Bob had a unique specialty. Known affectionately as Beetle Bob, he was a geologist specializing in insect remains found at post-glacial sites. The identification of the insects could indicate what the environment was like at the time of their existence since insects of that time period still live today, some found only in

71

particular habitats. We also welcomed Dinah Crader from the University of Southern Maine as an observer because of her experience in taphonomy—studying elephant die-offs in Africa.

Bruce and I provided an overview of the history of the tusk's discovery, complete with the 1959 film footage and a discussion of last autumn's field-work. Then we passed around the radar images of the pond. Finally we established a protocol for excavating the site specifically focused on bone recovery.

The afternoon session held in the museum auditorium brought together over twenty researchers. Greg Hart projected a video that he and I had created linking the 1959 film with news sequences from various outlets, followed by a slideshow of our work at the site and the screening of spoils at the Manchester lot. A detailed overview seemed obligatory for our specialized audience.

At that point we took a break so researchers could examine the mammoth bone fragments. Then the brainstorming began from our guests: Sieve above the marine clay every 5cm. for insect fossils; check for gnawing marks left by mice and shrews especially on long bones; look for marks caused by abrasion on rocks and sediment due to water transport; make a list of approved "observers"; a multi-authored paper should take publishing priority. Many more recommendations came our way but one set of comments I found compelling.

A geologist who worked on the site in 1959 admitted to the group that the dig had been disorganized. He congratulated us and then said that when a reporter asked if the tusk had come from an elephant killed in Maine in 1816, he and Dr. ___ baited him with the disinformation that tests showed it to be modern. No tests took place until we established its age in 1990.

Whatever happened in 1959 sealed any further interest in the Scarborough site at that time. I've often thought about that over the years. Had geologists taken more interest and examined what lay literally right under their noses, they would have made an historic discovery.

20. Hopes Dashed

Two weeks prior to our mammoth workshop, I received a fax from Dan Delea. A gridded map with shaded anomalous regions came through along with a message and a few sheets of raw data. The 300-megahertz dish had picked up what could be bones in the pond bed. The possibility that we viewed mammoth tantalized me. The previous year's success elevated my confidence of finding *in situ* bone.

On June 18th I went to the site to meet with a Portland television news crew and check on the progress of our logistics team, which had arrived four days earlier. Greg Hart had pumped out the pond and relocated a large snapping turtle and amphibians to the lower pond. Work platforms floated on the mud of the pond bottom, and up near a half finished kitchen that Scott Mosher was building hung a 500-gallon tank from a tower ten feet above the ground. To fill the tank, water would be drawn from the lower pool through a series of flexible plastic pipes. It would be chlorinated then used in a gravity-fed shower system directly beneath the water tank.

By the time we began our dig on June 28, everything seemed in place. All the hard work of planning, renegotiating with the landowners, and preparing the site for a summer of overnight camping lay behind us. Now the proof was in the pudding, and that pudding was the muck in the pond basin.

Bruce thought we might not need the use of heavy equipment so we began shoveling pond sediment from the floating platforms and hauling it away in wheelbarrows. But I began to doubt the wisdom of our excavation method and commented in my journal, "The pond bottom trembles like jelly when we work."

74

By the time noon break came we had hauled away many loads of sediment but mud slumped back into our hole erasing any sign of progress. Time to change tactics. JR agreed with me when I called and he assured me that a large backhoe excavator and dump truck would be on site the next day. Meanwhile we probed with no conclusive results.

Before the power shovel arrived the following morning, Judy Ritchie fell off the floating platform and sunk to her chest in mud. As we pulled her to safety she announced, "I swam with Hairy-it."

Judy Ritchie stands defiant shortly after being pulled from the mud. (Author photo.)

After cutting a level work platform into the gulley wall, the power shovel cleared mud from the northeast side of the pond bottom where Cash and Littlejohn had found the tusk. Work progressed slower than usual because of our fear that the shovel might damage undiscovered bone. Some anomalies on the radar graph were in that region.

By day's end signs of the 1959 dig appeared. The parallel marks of bucket teeth lay embedded in the clay, a signature distinctly different from our smooth-lipped shovel. I called Leonard Cash. When he arrived that evening, I hoped the claw marks might awaken a dormant memory, but he couldn't add anything to his previous story.

Over the next several days the shovel cut down to the marine clay layer on the north gulley wall leaving a shelf we could probe. As operator Dale Maxwell worked his way along the bank to create a stable access road for the shovel and dump truck, he dug into some of the remaining 1959 spoils still on the bank. Those truckloads were sent to a nearby sand pit where they would be screened with the equipment we had used the previous year. Once Dale reached a dozen feet beyond the pond's upper bounds, he began stepping down the excavation wall into two 5ft. high steps roughly 10 to 12 feet deep to meet OSHA approved safety standards. At the lowest step he scraped gravel from just above the marine clay and left the rest to us. Afterwards, he moved to the other side of the gulley and opened a long cut clear of the pond for geologists to study, while the rest of us formed a bucket brigade to clear spillage from step-down walls and the remaining gravel veneer over the marine clay bench.

The geologists were impressed with the size of the stratigraphic area we had revealed, and soon much debate about its interpretation flowed between Chris Dorion, Lisa Churchill, and Tom Weddle.

While Chris sampled the marine clay profile every 10cm. (4 inches) in depth, volunteer Mick Evans and I systematically pushed probes into the surface of the clay bench, hoping to make contact with bone.

On July 6 we returned from an Independence Day weekend to finish the clay-bench probing. We then extended the probing down into the pond basin and up the gulley another twenty feet on the clay-bench with no positive results. The radar anomalies I had been so hopeful about proved not to be bone. A skeletal reconstruction looked bleak.

A silt pond had been installed the previous year prior to the initial draining of the mammoth site. It consisted of a shallow depression dug into the gentle slope of a bank about fifty feet beyond the mammoth pond's outlet. Two layers of hay bales ringed the depression along with layers of sieve-like fiber material. The excavated dirt had been packed against the fiber layer ringing the inner bales. With our pump activated, water flowed from the mammoth pond through pipes that emptied into the shallow pit. As water filled the pit it seeped into the soil and through the packed dirt, silt fence fibers, and hay then joined the lower streambed cleaned of sediment. Because subterranean seepage provided a constant flow of water to the mammoth pond, every morning a sizable pool of murky liquid had to be pumped out.

As the day grew hot and humid, an official from the Department of Environmental Protection arrived to inspect the site and informed me that we would have to clean out our silt pond immediately to avoid contaminating the downstream waterway. I said we would get right on it, though I felt sure we could squeeze a few more weeks out of the system.

When television crews arrived by late morning, I fell into despair and I guess it showed on TV. That evening both landowners arrived at the site expressing sympathy about my disappointment. It became the high note of the day. We had worked through thorny issues, at times even getting lawyers involved, but that rocky road developed into a relationship that neither party wanted to abandon.

21. Little Things Count

Next day Bruce and Greg arrived with a small front-end loader and the big screen system we had used the previous year in Manchester. After unloading the screen at the nearby sand pit, Greg returned with the loader and dug sediment from our silt pond. Four of my faithful volunteers – Nancy Freese, Betty McWilliams, Judy Hunniwell, and Roger Willis – helped me clean out the remaining silt and replace the old hay bales with new ones. Then while Bruce studied the site, the rest of us went to the sand pit and screened until 3:00 p.m., when we had to quit because the temperature soared to near 100°F. I felt bad that Judy Hunniwell had to leave for home due to heat exhaustion.

The following morning the heat, humidity, and ozone climbed again toward the danger zone. I limited the crew working at the screen to a 12:00 noon shutoff. But even at that Judy Ritchie fell ill and left for home.

I worked that morning with Chris Dorion, Lisa Churchill, and Bob Lewis coring the marine clay bench. The tool used resembled a hand auger with a cylindrical shaft and a sharp cutting edge at its base. When the tool twisted into the clay, it cut a plug the length of the cylinder. After retrieving, a hinged door allowed it to be opened in half to remove the core, wrap it in plastic and send it to the University for analysis. Chris packed up the cores in his vehicle and left for Orono. From there he would travel to Aroostook County for a week of lake bottom coring. By 7:00 p.m. I hit the wall and crawled into bed.

*　　*　　*　　*　　*　　*

When I volunteered for the dinosaur dig in Montana, I experienced an environment so alien to me that it always filled me with wonder. It ceased to be a "bad land" in my eyes. The ever-present sagebrush, cactus, and mudstone opened to rich earth-tone panoramas that challenged my senses. Hoodoos embellished by the bronze light of late day shimmered against indigo canyons, like gilded sentinels guarding the sacred buttes above them; some sands in the near distance seemed so white, that only the dry oven heat convinced me they were not snow.

The author's on-site sketch of some of the survey party's work area.

Nor were the badlands devoid of wildlife. During a morning briefing, a sagebrush lizard climbed onto my index finger and sunned itself through the twenty-minute talk.

Sagebrush lizard on author's finger. (Photo unknown.)

Mule deer bounded across our paths. Pronghorns raced through open plains. Jackrabbits and wood rats scurried through the undergrowth. A four-foot-long prairie rattler in rich earthen hues flashed its black tongue six feet away, warning me to take another path.

Canyon north slope field sketch. East Montana badlands. (Author's illustration.)

Comparing my rich Montana experience with what my volunteers in Scarborough went through made me realize how dedicated my people were. They faced no beautiful vistas, only an expanse of slippery clay, gritty gravel and sand. The air – hot, steamy, and laced with ozone – left everyone drenched in sweat and exhausted at the end of the day. Volunteers ranged in age from nine-year old dinosaur enthusiast Ken Larson to Roger Willis, who celebrated his 80th birthday at the site. Besides Judy Ritchie another experienced and reliable volunteer who worked weekdays for the State of Maine Legislative Council joined my crew. His name was Patrick Norton. Pat had grown up in the

Portland area and had a special fondness for the old Portland Museum of Natural History. He had also earned a degree in paleontology at the Arizona State University and had considerable experience in excavating fossils in some of the western states. He became my stand-in on weekends. And that was greatly appreciated.

The only true reward for all my hard-working volunteers became the thrill of discovery, and fortunately most of them did find fragments of bone or shells important to our research. But even those that found nothing enjoyed the experience of having contributed to an historic event.

*   *   *   *   *   *

It surprised me to see a crew of ten people show up Friday, the morning after my long night's sleep from heat exhaustion because the weather had not changed. Concerned about the heat and ozone, I cut off work at the sand pit at noon and had the crew sift through the remnants of the 1959 spoils under the shade of pines until 3:00. Several of my volunteers returned on a daily basis through the oppressive heat. One of these was Roger Willis. I noted in my journal: "Roger is amazing – 80 years old, fleshed out and strong."

Everyone except Lisa Churchill and I left for the weekend by late afternoon. I again felt very tired and climbed into bed early, waking briefly later that night to listen to a coyote's serenade.

When morning came no volunteers showed, so I decided to sketch the exposure of the gulley's northeast bank, while Lisa sampled layers of clay.

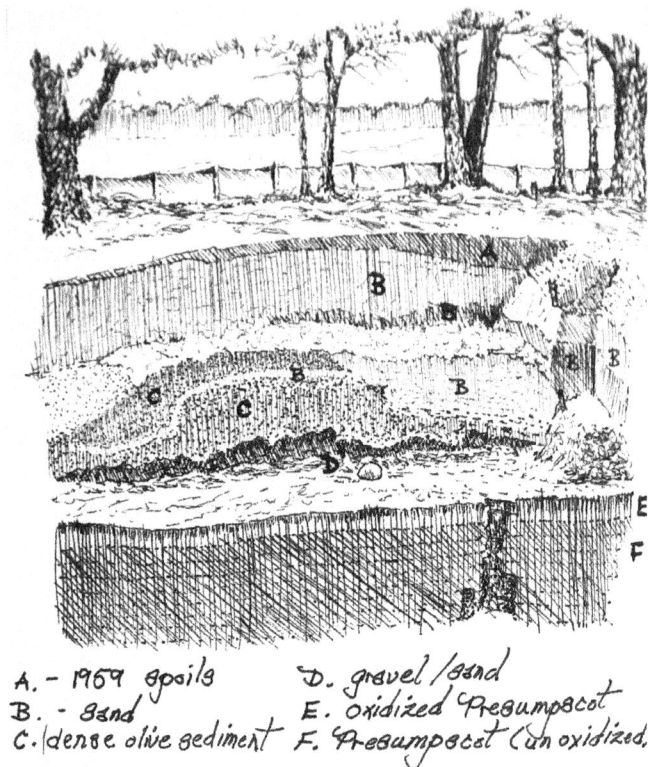

A. - 1969 spoils   D. gravel/sand
B. - sand   E. oxidized Presumpscot
C. dense olive sediment   F. Presumpscot (un oxidized.

Excavation site. (Author's illustration)

Moments later Lisa hollered to me, and I recorded her reason in my journal.

"Lisa has made an exciting discovery – an insect wing 2 cm.
below the interface between oxidized and unoxidized
Presumpscot (marine clay). We may not have mammoth
bones but I think we have something important."

It proved to be a key to the type of environment the mammoth, Hairy-it, lived in.

22. Personal Joy Professional Insight

Over the following days shovel operator Dale Maxwell showed his talent. On the campsite side of the gulley he carved a roadway down to the upper shore of the pond where a delta of sand had infiltrated. He laid a small section of corduroy road on the delta then settled his track rig on the platform of logs and began cutting down the upper far bank to extend our stratigraphic profile. In the process the shovel lifted the 1959 spoils' overspill from the crest of the gulley and dropped it in the trucks for later screening at the sand pit. Dale stopped at one point and plucked an eight-inch piece of rib from the spoils. Lisa, monitoring from above the shovel cut, also found a rib fragment, but we couldn't be certain the two sections came from the same bone because Littlejohn's bulldozer had scattered fragments as he cleared away debris in 1959.

By the time Dale finished stepping down the wall and revealing more of the marine clay for probing, he had exposed all of Leonard Cash's shovel marks on the north side of the gulley: the sign of another shovel "artist."

I called Cash and he arrived by mid-morning the next day, while Dale exposed the gulley wall closest to our campsite. Looking over his old excavation, he mentioned a new piece of information to me. After removing the tusk from the clay, Littlejohn insisted they replace it until the geologists arrived. The "geologists" Cash mentioned were likely associated with Christopher Packard's first trip to the site, but I could not confirm that because Packard died just prior to our first year's investigation, and I had ceased communicating with Dr. ___.

Bob Nelson and his assistant sample for insects at the Scarborough site. (Author photo)

Dale completed the stepped-back soil profile of the entire pond by 2:00PM on July 14. Over 60 feet of the marine clay on either side of the 30 foot pond was now exposed, only a third of which had been probed. With the heavy equipment gone, more people could probe the clay surface without concern but not for a few days. It rained most of that night leaving the new exposure too wet and slippery to safely work. We resolved to sift through a pile of spoils I had avoided because of its dense consistency similar to the mudstone I had encountered in Montana. The rain had softened it a bit, but it still became an all-day challenge to get through the small pile, which yielded no bone.

At the close of the day, Dan Fisher and his family arrived as expected. Dan and I had been in communication throughout the dig, and now that the clay around the pond had been exposed, he was eager to evaluate the site. Judy Ritchie and I talked science with him until 9:00 that evening.

The following morning I called the museum and Linda Carrell answered and told me to call home immediately. When I reached my wife Jeanne, she gave me fabulous news. We were now grandparents. Our daughter-in-law, Nici, had given birth without incident to a healthy, 8lb. 6oz. girl: Cassandra Estelle Hoyle.

Grampa and Cassie

I felt overjoyed, but the timing was against my visiting until the next day (Saturday) in the afternoon. Meanwhile, I needed to utilize Dan's knowledge and observational skills.

In order for Dan to study the geology of both sides of the gulley, we had to clean off the face of Dale's latest excavation. In so doing Dan discovered small depressions in the upper layer of marine clay where it color-shifted from olive to gray.

They were molds left by bivalve mollusk shells that had dissolved away, giving us objective proof that William Littlejohn and Leonard Cash had given me an accurate depth for the tusk's *in situ* location and that Dr. __'s field sketch was in error. Had the tusk been much higher in the clay, it would have shown signs of erosion from the organic acids percolating down from decaying vegetation and staining the upper layer of clay an olive brown which was the layer where Dr. __'s sketch placed the tusk.

Author's rendering of Dr. __'s field sketch, placing the tusk in brown clay—oxidized (acidic) zone. A rib fragment noted in the drawing as being found in "grey clay" should have been a clue for researchers in the 1959 investigation of the site that the tusk had been originally found within the marine clay—unoxidized (non-acidic zone).

23. The Haunting Tricks of Hairy-it

Unlike the human skull, dominated by hard cortical bone, an elephant's skull is composed primarily of a honeycomb structure that makes a ridged but lighter-weight skull with a thin cortical bone overlay. During our first year working the site we recovered over 300 bone fragments, mostly of the skull's honeycomb framework. I hoped the second year to find whole bones and enough to create at least a partially articulated skeleton, much as Christopher Packard had wished in 1959. Ground penetrating radar hinted at that, but it proved to be misleading.

The little pond in the Scarborough woods spawned abundant theories. The modern elephant grave had been proven false by radiocarbon dating. The mired mastodon was a mammoth, and no evidence had been found of disturbed sediment to indicate mire. A paleo-American meat storage site seemed unlikely since the soil profile showed no evidence of a bog. The kill site by a large carnivore would have left scattered bone at the same ground level as the tusk. But as we probed and peeled away layers from the marine clay, no support for that theory came to light. Judy Ritchie even got a note from someone who had a vision that we should investigate an area higher up the soil profile. I found it odd, amusing, and a little disturbing that so many volunteers embraced the idea since I had told them many times that no bone could have survived thousands of years in that acidic soil. We lost several volunteers to half a day of that nonsense.

Were we overlooking a clue?

I got a call from Dan Fisher one day while I probed the pond bottom. He and his family were heading to Nova Scotia from Cape Cod to see the recent discovery of an immature mastodon unearthed by our friends Bob Grantham and Kelly Kosera. He had dropped by the Maine State Museum to see our previous year's excavated bone, and noticed something odd.

> "He believes that a light blemish which he noted on one of our mammoth rib fragments…was caused by the basal plate of a barnacle," I noted in my journal.

Before our dig began Hal Borns had posed the question, "How did an elephant get in the clay?" Now the question was more complex. Every discovery provoked more questions, but only one burned inside me. Where could the *in situ* bones be? The obsession kept my mind constantly shifting to new locations. Sometimes I even fantasized them being where I had warned my volunteers they couldn't be. The weeks of living in the woods and working in the heat, high humidity, ozone, and periods of rain were wearing on me. I needed time to think things through logically, and I took it on Friday, July 30th.

That morning I sent the team to the log yard to screen spoils, while I stayed in camp to think through our find and record the following.

> "We have fragments of floating ribs, one vertebra (fragment), ribs (fragments), skull (fragments) and of course a tusk. Just about all these elements including the tusk are from the left side of the animal. Furthermore, Leonard Cash says that the 1959 photograph was staged to show the tusk in the position it was found. That position indicates to me that the inward spiral of the tusk was downward, thus, I think the animal was lying on its right side. We must dig deeper."

By late afternoon of the following Tuesday we finished screening the last of 30 truckloads of spoils at the log yard. From all those loads we managed to find only one small piece of rib. With that phase finished, we had a reason to celebrate, and 80-year-old Roger Willis capped the day off by serving a scrumptious dinner of sweet and sour chicken on wild and long grain rice.

24. A Bone to Pick

With the dawning of the first Wednesday in August we still faced the screening of several truckloads in the sand pit, and careful excavation of a veneer of spoils left on the northeast rim of the gulley. Some museum staff worked on a fiberglass peel of the soil profile at the sand pit. Dale had cut the wall to give geologists a better understanding of the mammoth site's relationship to the overall geology of the area. We planned to use it in an exhibit on the mammoth as a backdrop to the discoveries we had made. I wanted to include a peel of one part of the mammoth site that I had kept covered with a tarp since the dream incident that led people to excavate the sterile sand beds. The soil profile there consisted of sand layers and a dark organic region. The irony that it looked very much like a Paleolithic cave painting of a mammoth made it a great exhibit backdrop. Unfortunately, the wall beaded with condensation and could not be peeled. We resorted to photography.

Note the dark outline of what appears to be the drawing of an elephant head in this stratigraphic profile. (Courtesy of the Maine State Museum)

92

Throughout the summer, fragments of bone appeared from the spoils around the pond especially on the east end of the north bank where a broad thin layer of the 1959 debris persisted into August. I had discussed with Lisa and Dan my idea that Hairy-it might be deeper in the clay than we thought. Both considered it a reasonable assumption. But I did not want heavy equipment working at the site again until the remaining spoils had been examined. So on the fifth of August I assigned the team to concentrate on spoils of the northeast aspect, while I left for a noonday interview at a Portland television station.

"Great excitement at the site when I return. Donny (Basset) had unearthed a fragment of the mandible (we think). At 9" by about 5" it is the largest fragment we have yet found."

Don held the position of Museum Art Director, but on site he worked with Museum Workshop Supervisor, Scott Mosher, on the fiberglass peel at the sand pit.

"But just before 5:00 PM Kristina Oliveri (MSM Visitor Service staffer) found a small knob of bone protruding from the veneer of spoils only a few inches from Donny's mandible fragment which we left in place for the press photographers. Then as Kris exposed more bone I speculated that it was a pelvic frag. Linda Carrell helped in the excavation. 'It's a vert!' she said. As they cleared away dirt from the bone's processes I cried out. 'It's the atlas!!'"

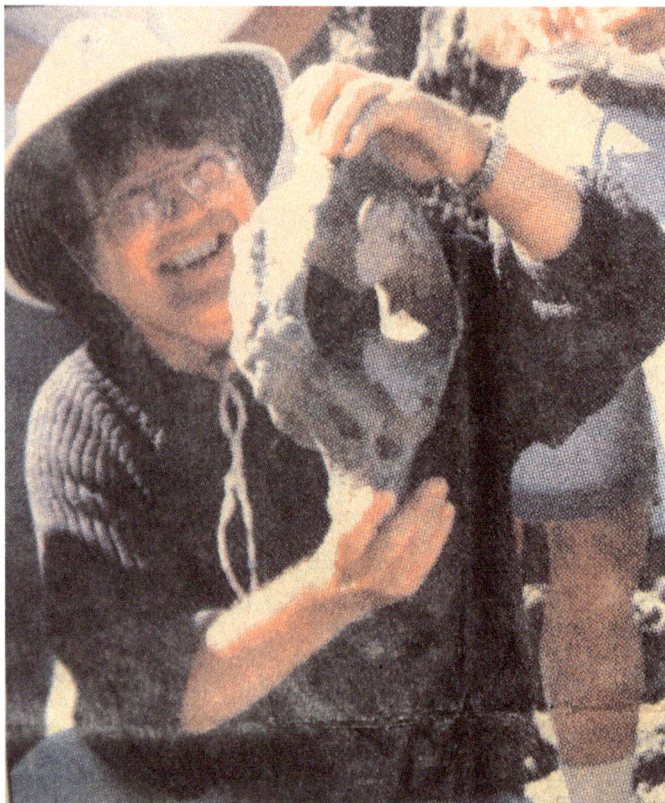

The author holding the atlas bone of Hairy-it. (Courtesy of the Maine State Museum.)

At last we found one whole vertebra, the bone of the spinal column that attaches to the skull—about 10 inches across. It is about three inches wide in humans.[1]

The next day Roger Willis arrived with the *Portland Press Herald*. We hit front-page news, and as a result twenty volunteers appeared at the site.

It was Friday and by evening everyone left except me and a foreign exchange student from Spain by the name of Miguel. He and I spent the evening talking about science, archeology, and the need for an individual's cultural identity. He spent several days and nights at the site before heading back to Spain. During that time he became one of the few people to find bone. Several years later I heard from this smart, engaging young man. At that time he was researching the history of the circus in Europe.

By August 27th we had completed all screening and examination of spoils. The power shovel had peeled away six-inch layers of clay from all around the pond and four feet into its basin without finding bone. The original 20 by 30 feet pond had become a hole 60 by 120 feet. The restoration of the pond would be dictated by the Department of Environmental Protection, but my responsibility at the site was over.

The final report in my journal read:

"It felt good to close the gate for the last time. After nine weeks of living in the woods the field work was over. Even though we did not find what we had hoped, the project was a success. Some good science will come out of this, and we have the makings of a dynamite exhibit. The temperature was 96, but I could see a few red maple leaves and a branch of golden birch leaves while I drove up the dusty logging road. Cooler weather would be coming soon. As I drove by the landowners' house, I breathed a sigh of relief. I turned the corner and hit the accelerator. Good-bye Scarborough. Hello Jeanne. How about a vacation?"

25. To Cut or Not to Cut

Several months after the close of our excavation, Dan Fisher arrived at the Maine State Museum. He, Hal, and I had been communicating on a regular basis about the project. Pairing the geological work done at the mammoth site with research done on the broader landscape of southern Maine showed that there might be a problem with our radiocarbon dates. We had found no indication of disturbed sediment in the soil profile to indicate a wallow. Furthermore, the description of the tusk's discovery from Littlejohn, Cash, and Packard, supported the view that the tusk had been found in a thin sand lens between layers of clay: a clear indication that it had been deposited with the sediment. Plus, Dan had found a blemish on the tusk that appeared to be from a barnacle. However, evidence from multiple sources showed that the time period for deposition of marine clay predated our radiocarbon dates by over a thousand years. We needed to date another sample, and Hal offered to fund the dating using a more accurate method that required only a fraction of material we had dated before.

Dan and I settled onto stools opposite one another at a corner of one of the examination tables in the Conservation Laboratory. With an electric bur tool we took turns drilling and supporting two significant fossils from the Scarborough site: the tusk and one of the molar teeth. We intended to drill deep enough into both skeletal elements to remove the least contaminated samples for radiocarbon dating. After an hour we succeeded in extracting ample material from the center of each fossil. These samples would be sent to a specialized laboratory along with fragments of the skull that would undergo greater extensive pretreatment before dating. From these three different samples we hoped that at least two of the radiocarbon dates would correlate. Then we met with Director JR Philips to discuss something I was hesitant about: the destructive analysis of our tusk.

Unlike molars, the modified incisor teeth of an elephant called tusks grow throughout the animal's life. Cones of ivory form from the vascular bed at the cupped end of the tusk, but variations in their deposition occur because of many factors, seasonal change being the most consistent one. The lean winter period creates a thinner ivory cone than times of plentiful fodder.

Imagine a stack of ice cream cones and you'll have a rough idea of how elephant tusks grow. The initial cone replaces milk teeth by the end of the first year of life. Adding cones from the bottom, your "tusk" lengthens, though it doesn't spiral or grow in girth like a real tusk. What it does do however is retain the cones of its "life" from the first to the last one you add. If an elephant's tusk is complete, it is the diary of that animal's life.

Having been schooled by Dan Fisher on the scientific importance of elephant tusks, I faced an agonizing decision. The only way to analyze the ivory layers required removing a two-inch-thick section of the tusk over its entire length. Being the Curator of Natural History I was responsible for the care of geological, paleontological, botanical, and zoological collections at the museum. The tusk had a unique importance as it was the only one ever found in Maine. To cut or not to cut? I was haunted by that question, which is why I wanted to discuss the pros and cons of it with Dan Fisher and JR Philips before committing to a decision.

In the end I felt comfortable making the decision to cut the tusk because the benefits outweighed the alternative. We could use the analyzed data from the ivory center section to tell the story of Hairy-it and integrate it into an exhibit with the remaining halves of the tusk. The cutting would be done at the Museum of Paleontology at the University of Michigan where the center slab would be archived for later analysis. But before that could be done, I insisted that a detailed plaster replica be made of the tusk.

26. Faking It

I first saw a natural history diorama when I was ten years old. My family visited my grandfather at his home in the Boston area at the time, and he took us to the Museum of Science. The mounted animals intrigued me, but the magic of frozen time intoxicated me. Here a three-dimensional snapshot of nature melted seamlessly into a painted view of the animals' greater landscape environment. I became hooked.

I started making clay figures of animals, both living and extinct.

Pre-teen author and his miniature nature scenes.

Eventually, I cleaned out a bookcase in my bedroom and built miniature scenes for my sculptures. As I got older, an air-dry clay that I could coat with watercolor paints became my preferred medium, and I became an avid reader of all things related to natural history museums including the exploits of people involved in collecting specimens for dioramas.

By the time I was 14 years old, I had created high enough quality sculptures to get the notice of a friend of the family who worked as an editor for a State of Maine publication. He knew Klir Beck, who was in the process of resurrecting the Maine State Museum that had been mothballed in state storage sheds since the late 1940s. A multi-talent, Beck had created full size and miniature wildlife exhibits that toured major cities on the east coast, celebrating Maine as a sportsmen's paradise. He had been granted the very large room that used to be the Maine State Museum in the lower south wing of the State House. But Beck's museum would not be the old cabinet style museum.

In the fall of 1959 he began creating four full-size wildlife dioramas, depicting animals native to Maine in each season. I met him during his construction of a rock ledge that would support a pair of black bears. Behind him hung plaster mushrooms and little painted bundles of dry grass suspended from clothespins. I felt so awestruck that I stammered when I started to talk. I brought the best examples of my sculptures to show him, and he spent considerable time examining them. He took special interest in a snapping turtle. Out of that visit, I would be invited to sculpt two life-size box turtles for one of the museum exhibits.

On a late winter day in 1960 at the age of 15, I presented my two sculpted turtles to Maine Governor John Reed and Curator Klir Beck in a ceremony at the Maine State Museum. The turtles would remain there until, fortuitously, I began my career in a newly constructed Maine State Museum in 1973. One of my early tasks required the removal of two crumbling turtle sculptures from the old museum. (From left to right: Gov. John Reed, Klir Beck, and author)

When I started work at the new museum, a slight, energetic man in his late fifties came in one day a week to work on the wildlife exhibits with museum technician Bob Barnard and me. His name was Fred Scherer and he had recently retired from the American Museum of Natural History in New York City. Fred behaved so unassuming that I didn't appreciate the range of his talent at first. But as I worked closely with him, his genius swept over me. He soon became my mentor, guiding me through every aspect of exhibit work, from building ground contours and rock ledges to fabricating leaves, flowering plants, and grasses. Beyond the nature exhibits Fred worked with me on stylized background paintings, detailed window views and the molding and casting of a wide assortment of exhibit elements. Some weekends Fred and I went out painting the landscape near his home in Friendship, and he would tell me about his experiences painting diorama backgrounds with famed artist James Perry Wilson. I cherished my times with Fred, but at the Maine State Museum they ended after ten years. The administration did not renew his contract and I never knew why.

Because of working all those years with Fred, I had honed my skills to the point that I felt confident that I could create a replica of the Scarborough tusk that could pass as the genuine thing from a viewer's distance. In 1994 I put my attempt on exhibit and it was a success. We did of course label it a replica.

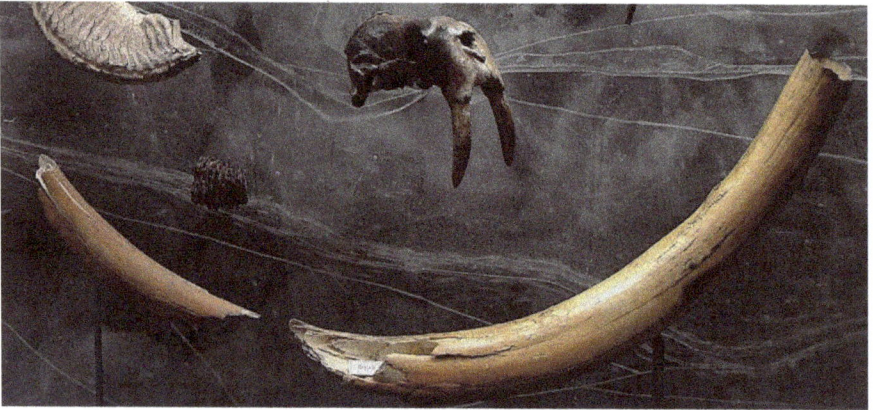

Replica tusk on display at the Maine State Museum. (Photo by author)

27. The Diary of Hairy-it

One day while working on the tusk replica, I got a call from Hal Borns. He was sending along a copy of our recent radiocarbon test results, but he wanted to share the news first. The chemical fractions tested from the sample deep in the molar and the extensively cleaned skull fragments hovered around a common date of 12,200 radiocarbon years before present. How that aligned with research done on the broader landscape of Maine would be the next question to answer.

Radiocarbon dating is not the same as calendar dating. In the early years of using C^{14} as a measuring tool it was assumed that radioactive carbon was produced at a constant rate as cosmic rays from outer space collided with nitrogen atoms in the atmosphere. It looked like a simple relationship could be drawn between radiocarbon dates and calendar dates. Tests began on historical objects with confirmed calendar dates. When found that radiocarbon dates varied from a simple relationship with the historical record, researchers began investigating ways of calibrating radiocarbon years to actual years.

Thanks to extensive studies of modern tree rings and those in the timbers of ancient buildings along with the analysis of annually deposited sediment layers from lakes and bogs, scientists have a far more accurate picture of the relationship between radiocarbon dates and actual dates. It has been found that the abundance of cosmic rays striking the atmosphere has fluctuated over time, leading researchers to discover that some objects are much older than their radiocarbon dates.[1] That was true for the remains of Hairy-it.

Chapter 27

Our radiocarbon date of 12,200 years calibrated to a little over 13,800 calendar years ago. That put our mammoth in a time period when the clay beds formed, another confirmation of the Littlejohn-Cash-Packard story.

* * * * * *

A few months after the replica of the tusk went on exhibit, I stood in a small lab at the University of Michigan watching Dan Fisher guide the blade of a band saw through the length of Hairy-it's tusk. A problem surfaced that we hadn't anticipated. The hardness of the resin we used to consolidate the ivory inhibited our progress, requiring periodic pauses to wait for the blade to cool. During those times I got an appreciation of tusk anatomy.

Dan showed me several examples of sliced tusk in his museum's specimen collection. The most obvious differences in thickness of the ivory cone layers were seasonal markers, while finer bands were progressively month, half month, and week events. Under the microscope the sight of day/night variations convinced me that locked in an unbroken tusk lay a detailed, unique history. Although our tusk had lost considerable ivory, Dan felt quite certain that enough of each ivory cone remained to analyze Hairy-it from childhood to death.

As I mentioned in Chapter 8, Dan was initially trained as an invertebrate paleontologist and had used an accepted method for studying growth rates in bivalve mollusks, which involved grinding samples from the growth ridges of shells and testing the powder for variations in particular stable isotopes. When Dan became interested in extinct elephants he applied the technique to the study of tusks. The growth bands most obvious in sectioned tusks appeared to show a pattern of seasonality. To test that hunch, Dan removed powdered ivory from each of those regions and tested for the ratio of stable isotopes of oxygen.

The most common isotope of oxygen is O^{16}, which is stable and has 8 protons and 8 neutrons in the nucleus of its atom, giving it an atomic weight of 16. Isotope O^{18} with two additional neutrons is also stable. Both of these isotopes are commonly found in water, but O^{16} water molecules are predominant. In fast moving rivers and streams these two types of water molecules become well mixed, but when the rate of flow subsides, the heavier water molecules tend to sink deeper into the streambed.

With that idea in mind, Dan reasoned that if an ancient elephant drank from a glacial river during the warm months when meltwater raged, more O^{18} would enter the elephant's system and be metabolized into the tusk's ivory than during the colder months, when O^{18} water molecules were more concentrated in the deeper regions of the slackened flow. This is just what Dan's analysis confirmed, giving him a seasonal framework in which to study the finer ivory layers. It was the first step in decoding an elephant's diary.

When our tusk was cut lengthwise the annual ivory cones took on the appearance of repetitive "V"s marking off the seasons of Hairy-it's life, similar to the following illustration. The darker regions of ivory are due to a concentration of thinner winter bands.

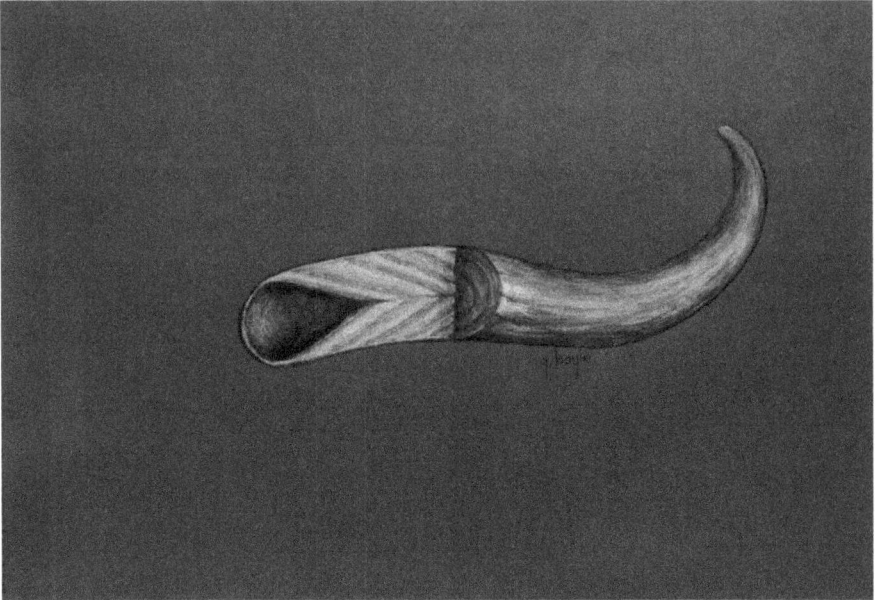

Tusk with cut away section showing growth layers of ivory. (Author's illustration)

By microscopically extracting samples from these annual layers, Dan could test for a range of chemicals locked in the ivory of individual cones and link them to seasonal changes and particular events in the mammoth's life. Techniques by then had been developed to determine in female tusks the number of pregnancies, because fetal development drawing away calcium for bone growth resulted in thinner annual ivory cones. Other analyses could reveal nutritional stress, feeding patterns, migration events, some health issues, time of death and more, as techniques developed to "read the diary."

28. The Ghost of Old Bet

When I began investigating the Scarborough tusk's provenance and identity, the name Old Bet kept surfacing. I knew of her only as the elephant killed in Alfred, Maine, sometime in the 19th century. But with every mention of her name, I became more curious about her and why some people believed she lay buried in Scarborough, sixteen miles from where she was shot.

Assuming she performed in a traveling show, I scanned several circus books in the state library but found them conflicting about her history. They did agree that she was important in the birth of the American circus. They also agreed that an irate farmer killed Old Bet at Alfred, Maine, in July of 1816.

I went to a microfilm reader and scrolled a film of the July 1816 *Portland Gazette*. In the July 30 edition I stopped on the headline, "DEATH OF THE ELEPHANT," with the following story:

On the 24th inst. about 6 o'clock in the afternoon, some unprincipalled villain screened by the bushes which skirted the road where the Elephant was passing in the town of Alfred, in this county, accompanied by the persons who had charge of her and fifteen or twenty citizens, discharged a musket loaded with two balls, which both entered the body a little back of the shoulder bone within two to three inches of each other – The Elephant after traveling a few rods fell and expired – Endeavors have been made to discover the author of this base act, but we have not learned that they have proved successful – the balls it is said passed within five or six feet of some of the persons in company. Great grief was exhibited by the persons who had charge of her and particularly by the mulatto man, who had long discovered great affection for her. The barbarity of the deed produced a sensation in the public mind, correspondent to its enormity and it is our ardent wish that punishment may in this case, tread close upon the heels of transgression.

The primary account had a vivid quality lacking in the books that I'd read. Even the melodramatic parts seemed quaintly appealing. I wanted to read more primary sources, but for the present I had to concentrate on the account that led to rumors about Old Bet at the Scarborough site.

On September 5, 1959, while geologists probed the clay, staff reporter Waldo Pray's article appeared in the *Portland Press Herald* questioning the identity of the tusk's owner. He had just read about Old Bet and immediately thought her bones might be those in the North Scarborough gulley. So he posed the question, "Are geologists wallowing in the muck for the remains of a prehistoric mastodon or are they desecrating the last resting place of a lovable old circus elephant named Bet?"

It was a good question since no trained geologist had been present during the tusk's removal from the clay and subsequently could not debate details of its *in situ* environment. But Pray followed up his question with some shaky ideas that he called evidence.

Best evidence available also indicates that the musket – not reputed for its penetrating power – merely infuriated the poor beast and that she went on a cross country rampage.

Evidence indicates that the rampage ended when she got mired in a pit of clay and sand, eventually sinking out of sight.

Bet would have had to travel sixteen miles in a straight line direction to get to the Scarborough site, and it is very unlikely that a raging animal would travel in a straight line that far. Furthermore, if she survived the shots and became enraged, a posse of town folk would probably have gunned her down before she could get far from Alfred.

Before publishing his story, Pray should have discussed his idea with a geologist and perhaps gone to the site. Leonard Cash had to remove a thick layer of bedded sand before encountering clay. No modern day wallow existed at the site, and it was an alien environment for quick sand.

However, the newspaper article became the perfect foil for Dr. __ and his associate at the time. No one had money for radiocarbon testing to date the tusk.

On September 10th Pray updated his article with a more reasoned defense for his argument. By then he had been in contact with geologists and Christopher Packard and may have seen a few primary sources. Packard had found a newspaper advertisement for a circus appearance in Portland on August 8, 1816. Pray latched on to the idea that Bet's remains might have been such a curiosity that they were hauled to Portland for exhibition before being dumped in the Scarborough gulley. But that notion strained credulity when he read claims that the stuffed elephant had been exhibited in the American Museum in New York. Was Bet skinned and reassembled in New York? Pray called the American Museum of Natural History and spoke to a collection manager who became aghast that one of their elephants might be missing. The institution Pray needed to contact hadn't existed for nearly a hundred years.

One thing in Pray's article I found particularly intriguing was that Packard had changed his original views about the tusk.

> Packard said the possibility that bones discovered in a
> Scarborough field are from a walrus or a mastodon has
> now been definitely eliminated.
> "On the elephant versus mammoth theory, we're first
> going to do all the historical searching we can," the society
> director declared.[1]

What gave him the confidence to change his view regarding the mastodon? No diagnostic bones had been found. On the other hand, despite the damage done by the power shovel, the tusk must have been in remarkable condition and its tight spiral more obvious. With Packard dead the answer appears lost to history.

* * * * * *

Limited resources were available for me to research Old Bet at the Maine State Library in the 1990s, but Chief Curator Edwin Churchill told me about a collection of early 19th century newspapers at the Hubbard Free Library in the little riverside city of Hallowell, not far from the Maine State Museum.

Upon my arrival at this small, but exquisitely ornate library, the librarian directed me to the historical collections where folios of a local 19th century newspaper, the *American Advocate*, resided along with another local newspaper, the *Hallowell Gazette*. Locating the 1816 editions, I turned to the issue following July 24, the day the elephant died. The same description I had seen on microfilm of the animal's death appeared. Hoping that there might be information about Old Bet's burial, I continued my search. In the Saturday, August 3rd issue an extract of a letter appeared dated 26 July from Alfred, York County.

"The late keepers of the Elephant have arrived in Boston, with her skin and bones…"

The proof at last! Only the internal organs and meat of Old Bet had been buried in Maine, and in a rural setting of hardships, it's unlikely all her meat went to waste.

While doing more research at the Hubbard Library, I discussed my project with the librarian who immediately asked if I would give a talk on Hairy-it and Old Bet. I agreed and in 1994 spoke before about 30 people at an upper room in Hallowell City Hall.

Much of my talk centered on the dig we had done, then I concluded with the fate of Old Bet.

Few people asked questions, but one older gentleman peppered me for information. His name was Waldo Pray, the same reporter who had brought Old Bet to Christopher Packard's attention back in 1959. He stayed and talked with me awhile. Toward the end of our conversation he told about dropping by the office of one of the geologists some time after the tusk's discovery. That researcher told him that radiocarbon dating had proved Pray right. The tusk had come from a modern elephant. That geologist was Dr. __'s associate. Pray's forehead furrowed as he said, "I guess it's time to bury Old Bet."

Watching him walk away, heaviness settled over me. A person who had based his career on objective truth had led Waldo Pray astray, and now he knew it.

29. Shunning the "Dark Continent"

On November 2, 1795, twenty-five-year-old Jacob Crowninshield, commander of the commercial vessel *America,* sat at a desk in Bengal with pen in hand. Having been raised in his father's shipping company, he had experienced over a decade's work on the docks and out at sea. George Crowninshield and Sons had grown to be one of the largest firms in Salem, Massachusetts harbor, thanks to the European and West Indies trades. Now the company took the risk of extending its reach beyond Africa. The Jay Treaty, which gave the United States and Britain limited favored nation status with each other, cooled some of the tensions between them that had built up since the conclusion of the American Revolutionary War. The treaty had been signed on November 19, 1794, and was ratified by the U. S. Senate six months later.[1] The Crowninshields wanted to be strategically located to take advantage of trade with India (then named Bengal) as soon as the treaty allowed.

Jacob arrived in Bengal with his younger brother, Benjamin. At that time their early trade likely involved a certain amount of mercantile reconnaissance to understand local businesses and secure relationships. Their brothers, George, Jr. and John, would arrive later on the company's vessel *Belisarius* to take advantage of Jacob and Benjamin's inroads. By then Jacob and Benjamin would be headed home.

America would be raising anchor in Calcutta harbor on December 3, 1795, but the impulse to describe Jacob's recent purchase to George and John had grown too great for him to wait another day to write the letter, though it would linger in Calcutta unopened for three months. He fingered the locks of his dark hair then slid his hand over a high cheekbone to cup his chin. His lips parted

112

in an open smile that brought a boyish charm to his aristocratic features as he wrote. "We take home a fine young elephant two years old, at $450.00…I dare say we shall get it home safe, if so it will bring at least $5000.00." Then an impish taunt, directed at Benjamin, flowed from his pen. "I suppose you will laugh at this scheme…This was my plan. Ben did not come into it, so if it succeeds, I ought to have the whole credit and honor too; of course you know it will be a great thing to carry the first elephant to America." [2]

Not long after the vessel *America* docked in New York Harbor on April 13, 1796, Jacob Crowninshield exhibited his elephant in the Bowery outside the Bull's Head Tavern [3], where he sold it to Mr. John Owen for $10,000.00. That handbill of sale resides today in the collection of the American Antiquarian Society.

Owen put the elephant on tour along the eastern seaboard, even exhibiting her at the Harvard Commencement of 1796.[4] He then wintered in the South to acclimate the pachyderm to the New World climate.[5] By the time he reached Salem in August of 1797, people from great distances came to see John Owen's fabulous beast. August 30th Reverend William Bentley joined a crowd in the Market House to get a glimpse of the elephant. Fortunately, he got close enough to record details in his diary.

> His tusks were just to be seen beyond the flesh… We say
> his because (of) *sic* the common language. It is a female
> and teats appeared just behind the fore-legs.

With Bentley's confirmation that the elephant was female it has been assumed by many writers that this animal would acquire the name Old Bet.

My earliest investigations through recent books and articles in magazines all agreed that Old Bet somehow launched the modern circus, but I soon realized that a good deal of contradictory speculation ran through many of these works. Old Bet had become a legend and like most legends, half-truths and exaggerations distorted her real story. One disagreement centered on the date of her arrival in America. 1796 stood out as the favored date because the Crowninshield elephant had been so well documented, but other authors suggested dates of 1804, 1805, 1807, 1815, 1817, 1823 and 1826. Having already found a primary reference to Old Bet's death in 1816, I considered the next favored date in publications, 1815, but I found no primary sources to support that date. Then in the Maine State Library I located the following advertisement for Old Bet in a May 1816 edition of the Eastern Argus newspaper.

NOW OR NEVER!! A FEMALE ELEPHANT TO be seen at Timothy Boston's COLUMBIAN TAVERN, in Portland, from the 29th inst. Until the 4th of June…She is **15 years old**…" (My emphasis)

Was this Old Bet? If the 1816 advertisement was accurate about this elephant's age, it could not have been the Crowninshield specimen because that elephant would have been twenty-two years old.

Looking for clarity I continued my paper search until I found solid evidence of Old Bet in the December 31, 1821, *New York Post* which read, "…the Elephant, **which was imported into America in 1804,** and after being viewed with wonder and astonishment, by our fellow citizens, for about 12 years, was killed by a barbarian, in the town of Alfred, County of York, and State of Maine, 24th July 1816." (My emphasis)

Proof at last that the Crowninshield elephant was not Old Bet.

With further investigation I found the late Stuart Thayer, a respected circus historian, cleared up the confusion about Old Bet's arrival in 1987 by finding in an early Boston newspaper that "[t]he second elephant to reach these shores arrived **June 25, 1804,** and was a four-year-old female from Africa … Edward Savage, a Boston artist, exhibited her…"[6] (My emphasis)

Prior to Thayer's discovery most, if not all, historians believed Bet had arrived from India where the Asian elephant had a long history as a work animal. Under those conditions calves were forcibly weaned when two years old, the age at which Crowninshield purchased his elephant. Bet landed at the docks of Boston Harbor at the age of four years, the natural weaning age for an elephant.

Before Old Bet came into Boston Harbor, a number of newspapers advertised for her sale even in New York City. The following ad appeared in the *Evening Post* (New York City) from May 30 to June 20,1804.

A RARE CHANCE FOR SPECULATION

For Sale a Female Elephant, lately imported from East Indies.
She is about **4 years old**…and is attended by a keeper, a native of Bengal, who will be of great service to the purchaser.
Apply at the sign of the Rising Sun, Marlborough – street, Boston. (Emphasis mine).

The avoidance of the true African origin in the ad may be due to confusion with the Crowninshield elephant because it had been exhibited in New York and Boston in past years, but I suspect an intentional reason for masking its true origin. People of the early 19th century viewed India far more favorably than Africa, especially citizens of the northern U. S. where most states had abolished slavery by the end of the Revolutionary War. A forceful movement swept the remaining states in that region to follow suit, creating an atmosphere that made most folks want nothing to do with a land acting as the source of the slave trade. Circumstances seemed to dictate that an Indian elephant could be sold at a much higher price than one coming from "the Dark Continent." In one account of the elephant's death, the keeper, whom the advertisement describes as "a native of Bengal," was likely African, as it states that it "would have melted the heart of the most obdurate, to have beheld the agony of grief and despair which the poor **black**, the Elephant's attendant, manifested…" My emphasis denotes perhaps the more accurate description of the keeper than in the advertisement. Although when the Crowninshield elephant began touring, at its stop in Rhode Island newspapers warned:

> The Public are hereby cautioned against trusting William, the
>
> **black** Man attending the Elephant, as the Proprietor will not
>
> pay any Debt of his contracting.[7] (Emphasis mine)

Crowninshield did deliver a trainer from Bengal with his elephant, but was he given the name William? It is a mystery as to whether the original handler traveled with John Owen, or an African.

The 1804 ad contributed to the distortion of Old Bet's history and a later debate over whether she had tusks, common in females of African species but rare in the Asian form. Reverend Bentley's remark that the Crowninshield elephant's "tusks were just to be seen beyond the flesh" is an accurate description of most female Asian elephants.

30. What About Hairy-it

Hairy-it the mammoth presented her own identity crisis.

After Linda Carrell and her team finished cleaning the nearly 400 skeletal fragments, Dan Fisher spent time in the Conservation Laboratory of the Maine State Museum methodically examining critical parts of the animal. Many of the smaller pieces came from the honeycomb-like matrix of the skull. Dan determined, as we suspected earlier, that most if not all of the remains came from the left side of the mammoth. He took precise measurements of the molar teeth, made a detailed note of a scar on one of the rib fragments, then requested closeup photography of the small bleached smear that he suspected was caused by the basal plate of a barnacle on another rib fragment (noted in Chapter 23).

Several weeks after Dan received the photograph he requested, he called me to say he compared the bleached smear on the rib with a series of barnacle scars. The scars and smear matched. Dan determined that the mammoth remains had been exposed to ocean water for at least two months before being buried in accumulating marine sediment that would become clay. He also told me that certain characteristics of the teeth did not conform to those of a woolly mammoth [*Mammuthus primigenius*].

*   *   *   *   *   *

About one and a half million years ago, a population of Siberian steppe mammoths [*Mammuthus trogontherii*] crossed the land bridge that existed during the Ice Age between Siberia and Alaska. Extending its range along an ice-free shoreline or interior corridor into the present day region of the United States, the steppe mammoth developed into a new species known as the Columbian mammoth [*Mammuthus columbi*], which roamed across the continent and as far south as Costa Rica.

At 13 feet at the shoulders, the Columbian mammoth grew considerably larger than the present day's largest elephant, the African savanna species. It was also more imposing than the woolly mammoth that evolved in western Asia and extended its range into North America about 400,000 years ago. By then the Columbian species had dominated the region for over a million years.

It is believed that the Columbian mammoth had a coating of hair less dense than its new neighbor because much of its range extended far south of the ice sheet in areas of open land with scattered tree growth. If so, it likely shunned regions close to the glacier except during the warmest months, when the nitrogen rich meltwater stimulated a lush growth in those grazing lands.

The southern glacial front sloped upward from its low melting profile, north to the vast body of the ice sheet where the thickness ranged one and a half to two miles. Air hugging this frozen plateau settled colder and denser than the air above it, and at the ice sheet's borders it slid downslope because of its weight. These katabatic winds ranged from mild to major hurricane force. Only a well-adapted animal survived most of the year on the grasslands near the windy glacial front, and that was the woolly mammoth.

Unlike its Columbian cousin, the wooly mammoth became more compactly built. At the height of a modern day Asian elephant and insulated with a subcutaneous layer of fat over three inches thick, plus a dense woolly coat covered with long guard hairs, its appearance must have been robust.

Author's scale model creation of a female woolly mammoth in its winter coat.

To minimize the risk of frostbite, its ears and tail were greatly reduced, and the trunk was covered with fur terminating in a finger-like projection that was likely used to pluck selected food delicacies. The tusks in both sexes grew more tightly spiraled than modern elephants and became massive in adult males. It is thought they were used to break ice from waterways and sweep snow when the mammoths foraged in winter. As with modern elephants, males used them in sparring for mates.

During the warming months of spring, woolly mammoths shed their winter coats and depleted much of their fat layers as their behavior shifted toward mating. At the same time Columbian mammoths moved north into the lush prairie-like environment and coexisted for a while with their new neighbors.

Cross breeding resulted and the hybrids showed characteristics of both species, even in their teeth. That created the combination of features Dan Fisher found in the molars of Hairy-it, though features of the woolly mammoth remained predominant.

Hairy-it appears to be an example of what had once been named the Jeffersonian mammoth [*Mammuthus jeffersonii*]. Thanks to recent research we now know this is a fertile hybrid, not a true species.

31. An Elephant, an Artist Showman, and a Wealthy Sea Captain

Menageries have a long history among the aristocracy in Europe. Their beginning stretches back to at least the Roman Empire when emperors kept wild beasts as a symbol of their power and magnificence. Most of these animals came from exotic locales, and large beasts of prey often became the preferred target.

Periodically, subjugates of the royal menageries became expendable entertainers in a bloody spectacle that came to its height in the Roman Coliseum where the emperor could express his dominance of nature to the citizenry on a grand horrifying scale. The frequency of such events depleted predator populations across Europe and the Middle East and contributed to the extinction of the Asian lion in those areas, requiring longer collecting forays. As time passed and more of the world opened to exploration, the discovery of vast varieties of unusual life forms impacted the menageries.

By the seventeenth century a wide variety of birds and small animals amused the gentry, expanding the size and content of the royal menageries. These animal "gardens," some of which had been periodically opened to the more refined members of the general public, expanded public entry and took on the role of celebrating the wonders of the creator's Great Chain of Being. This "Chain" was based on Aristotle's system of ranking components of the natural world. The Church embraced but modified this system by placing God as supreme and the angels then the king under him. The rest of mankind was eventually placed on a scale with white Europeans at the top.[1]

A century later with the rise of the Enlightenment, a more rational approach to nature and an increasingly educated citizenry precipitated the metamorphosis of many royal menageries into public institutions. In the early 1800s one author referred to them as, "among the most rational gratifications of curiosity."[2]

The menagerie's long aristocratic heritage and its more recent recognition of having intellectual value would have been a potent enticement for a young ambitious man to seek out something exotic. Such a man was John D. Sloat, who had been raised in a wealthy, slave-owning family and would become a commodore in the Navy and claim California for the United States.

On March 26, 1804, the schooner *Only Daughter*, mastered by Sloat, glided into New York Harbor loaded with coffee and cigars from Cuba. Sloat had served in the Navy until age nineteen when the 1800 Peace Establishment Act reduced military manpower. At that point he began to master commercial vessels out of New York Harbor. After leaving *Only Daughter* at the dock the twenty-two-year-old remained in port until May 9, 1805, when he mastered the square-rigger *Clarissa*.[3]

Flush with money and having plenty of time on his hands Sloat became aware of the advertised elephant and bought it.[4] The purchase likely took place in the early summer, sometime between June 20 (the elephant's last advertisement date in the *Evening Post*) and its arrival on June 25 in Boston Harbor. Sloat, however, may not have seen the elephant land at the dock, but documents confirm that well known artist and showman Edward Savage did.

Savage professed to be a self-taught artist though his skill in using perspective implied a trained hand. He did however struggle with anatomy as evidenced by some of his surviving portraits, but that did not hinder him from painting some of the dignitaries of his day. His round boyish face and affable charm attracted many friends, but his thirst for recognition made aristocrats his focus.

To promote his career in his late twenties, Savage contacted Joseph Willard, the president of Harvard College, and offered to do a painting of President George Washington that he would donate to the college if Willard would arrange for a sitting. Willard wrote the request to Washington on November 7, 1789.

Washington accepted. By then his early aggravation about posing had switched to pleasant compliance as he noted in an earlier letter about the subject.

"I am so hackneyed to the touches of the Painter's pencil, that
I am now altogether at their beck, and sit like Patience on a
monument, whilst they are delineating the lines of my face." [5]

The experience launched Savage's career for painting American dignitaries and linked him with other notable painters such as Charles Willson Peale, who had done several paintings of Washington and had created a successful natural history museum in Philadelphia by using his artwork as backgrounds for taxidermy specimens of his own creation. [6]

After spending three years in London studying engraving, publishing his own engravings of American dignitaries, and creating a panorama of that city, Savage moved to Philadelphia in 1795 where he exhibited the London panorama along with other artwork and a natural history museum that was likely inspired by Peale. Ten years later he and Peale exchanged natural history specimens.

In 1801 Savage settled in New York where he prepared to reopen the gallery he had established in Philadelphia. That same year midshipman John

Drake Sloat was discharged from the Navy on May 21st. Before Sloat returned to active duty in the War of 1812 he commanded over fifteen commercial vessels. Up until 1806 he captained ships out of New York Harbor that often sailed to exotic locales and brought back natural curiosities along with their standard cargo. These curiosities constituted the bread and butter of many 18th and 19th century natural history museums and, as such, became of interest to Edward Savage.

In 1802 Savage opened his Columbian Gallery of Painting and City Museum in New York and advertised himself as a "historical painter and museum proprietor."[7] To expand his natural history holdings, he had purchased a well-established museum collection, Baker's American Museum, which had been founded in 1791 by wealthy merchant and philanthropist John Pintard. In the original American Museum, Pintard functioned as secretary and Gardner Baker as keeper.[8] Four years after its founding Baker took over the museum and put his name on the marquee, but three years later he died. Then in 1800 his widow passed away and the museum went up for sale. Savage arrived in New York at the perfect time to buy Baker's museum.

Savage chose as his keeper John Scudder who had a passion for natural history and had an exceptional talent for taxidermy. Scudder worked as a serious student of the natural world, while Savage played the showman and concentrated on his paintings. Within eight years Scudder bought Savage's natural history collection and opened his own American Museum.

Edward Savage's talent as a showman may have precipitated his arrival in Boston Harbor in June 1804.

Whether he went there to purchase the elephant for himself, or acted as John Sloat's agent, is unknown. He did tour the elephant through New England and other areas of the northeast, but there is no indication that he arrived with the elephant in New York City where he lived at the time. Perhaps he left the elephant at the Drake estate in Goshen after his tour where slave labor could accommodate the husbandry demands of an elephant. Joseph and Amy Drake were a wealthy couple who had raised their grandson, John Drake Sloat, after his parents died. The elephant would next be shown the following year in the Hudson Valley, the neighborhood of the Drake estate.[9]

32. A Sea Captain, an Elephant, and a
Farmer Meet at a Tavern

During his 1918 Peekskill, New York, centennial address, U. S. Senator Chauncey Depew described the vital role of the early Hudson River sloops and how important their captains were to the river valley communities, noting that they "brought back from their trips to New York all the news of the day… the sloop captains not only carried the produce and cattle, but marketed them in New York, so that they were both navigators and commission merchants." [1]

That aspect of Peekskill's history equally applies to Ossining, ten miles south of Peekskill, which in its past was called Sing Sing. Of the many sloops transporting cattle from Sing Sing to New York City, one was owned by Hachaliah Bailey and his brother Stephen who came from Stephentown, about 15 miles northeast of Sing Sing. Stephentown, renamed Somers in 1808, became strategically located at the start of the 19th century due to two major roads intersecting in the rural town, a road carrying traffic from Peekskill and Danbury (Connecticut) and a road to Sing Sing. The Bailey home sat not far from the intersection, and Hachaliah took advantage of its location. Though brought up in a farming family, the lean, energetic young man developed much broader interests: the river sloop being one of them, managing road travelers another.

The Peekskill and Sing Sing roads became stagecoach routes and major arteries for the seasonal movement of cattle to New York. Traveling people needed places to stay along the way, and at the age of twenty-seven Hachaliah Bailey planned to provide those accommodations. In 1802 he received a license from the Excise Commissioners of Stephentown to keep an inn or tavern.[2] He likely intended to use the license for a structure linked to his home, where he could cater to cattle drovers and stagecoach travelers. Because he came from a large family, relatives took over the business while Hachaliah and Stephen drove the family cattle and those of some neighbors to Sing Sing and boarded the herds onto the Bailey sloop for New York City. After sailing down river about fifty miles they docked at the southern Bowery just outside the metropolis where stockyards clustered around the Bull's Head Tavern.

The Bull's Head held a long history as the watering hole for drovers and sea captains. It began when a prominent New Yorker named Adam Van der Burgh requested to lease a vacant farmhouse on Trinity Church property as a tavern. Church trustees agreed to the plan and in 1760 the Bull's Head opened for business at which time butchers, cattle drovers, sea captains, and various others began to gather for drinks and conversation in the smoky, dimly lit barroom or during fine weather under the shade of trees surrounding the tavern.[3]

By the time Hachaliah and Stephen began their sloop runs to the City, the Bull's Head had become famous as the establishment where George Washington commanded his troops and later refreshed before watching the final British retreat from New York Harbor. It had also grown into a robust center of commerce for butchers, cattlemen, and drovers. Cattle pens replaced trees and the air hung ripe from the decades' stench of slaughter. With the added temptations of alcohol, prostitution, and gambling, it proved to be a sink or swim testing ground for young entrepreneurs like Hachaliah Bailey.

Hachaliah succeeded at the Bull's Head challenge. Being raised in a farming family he came to the tavern with older relatives when at least in his early teens and learned the pitfalls of many drovers. By the summer of 1806, on the cusp of his thirty-second year, Hachaliah's business skills had become well honed. That year his fortune likely changed in a positively unexpected way.

Not so for John Sloat whose agony began in November 1805 when the schooner *Hunter*, which he commanded, left New York and grounded eight days later on a reef off Bermuda. By the time he returned to New York due to a series of detentions by the British, it was the summer of 1806.

That incident soured Sloat from shipping out of New York Harbor. From then on, he sailed from Baltimore, Maryland, beginning with the command of the 113-ton schooner *Vigilant* on August 4, 1806.[4] That route to the Caribbean brought vessels southwest of Bermuda. But prior to Sloat setting sail, his holdings in New York needed trimming, one of which was an elephant. And the most logical place to sell an elephant in the City was at the Bull's Head Tavern.

It would not be the first of its kind to be shown at the Bull's Head. In 1796 the Crowninshield elephant had been displayed there shortly after its arrival, and may have been where John Owen purchased the animal. Both a teenaged John Sloat and an early twenties Hachaliah Bailey may have seen the elephant at that time. Certainly they heard about the spectacular beast.

So in the summer of 1806 a sea captain, a farmer, and an elephant met at the Bull's Head Tavern, and a sale was made. John Sloat walked away with $1,000, while Hachaliah Bailey loaded an elephant onto his and brother Stephen's sloop for the journey upriver to Sing Sing.[5]

The 1806 purchase. (Author's illustration)

A tale later surfaced that Hachaliah bought the elephant to plow his fields, but that seems unlikely and may have been based on remarks said in jest at the Bull's Head. It doesn't make much economic sense to purchase an elephant for $1,000 only to plow a field. The maintenance of an elephant is not cheap now, and it probably never was, even with an experienced handler. Awareness of John Owen's success with the Crowninshield elephant must have been the primary motivator.

From the docks at Sing Sing Hachaliah avoided onlookers by walking the fifteen miles to Stephentown at night with the elephant and her handler. On reaching home he put the elephant and trainer up in his barn. Soon afterwards he named his elephant Bet or Betty, after his daughter Elizabeth Ann who had been born the previous year.[6]

After having the success of charging locals to see the curious acquisition on his property Hachaliah decided to tour the surrounding region with his elephant. "He started out to show Old Bet with a wagon of hay, a horse to draw it, and an assistant. The admission fee for an entire family was either a coin or a 2-gallon jug of rum."[7]

The assistant was of course the African handler Bailey acquired with the elephant. Whether he lived free or slave is unknown. In 1799 New York passed a gradual abolition law, but it is difficult to know how that applied to an elephant handler.[8]

33. Deciphering the Past with the Present

During the 1993 field season, our geological team of Tom Weddle, Chris Dorion, and Lisa Churchill periodically gathered at the Scarborough site to debate key issues of the exposure. Sometimes they sampled strategic levels of the marine clay. Lisa collected shells and recorded their location. Then the geologists left to search out other nearby locations to interlink their findings, sharing data with others researching the region, or archiving their samples at the Maine State Museum.

Guest geologists from various institutions arrived to offer their interpretation of the site's soil profile or, in one case, link the site to the broader landscape through GPS.

We knew early on that the mammoth site sat in a shallow tributary of the Nonesuch River, which, as the name implies, is not a significant waterway, but late one afternoon Chris Dorion arrived with fascinating news. Geologist J. M. Clinch and my old friend Woody Thompson at the Maine Geological Survey found that the original glacial river diverted just to the north of us from the Nonesuch Valley. Erosion caused it to become the main artery of today's Saco River. Before that the Nonesuch had been a swift river all the way back to the time of Hairy-it.

Lisa Churchill's collection of shells also told a story. Most of her specimens consisted of mollusks, such as snails and bivalves. They represented species that survive today, some limited to specific living conditions making them ideal indicators for understanding the ancient environment.

Chapter 33

The tiny clam *Portlandia arctica* mentioned in Chapter 12 is one of these indicator species found today in abundance only at the mouths of rivers discharging turbid meltwater from nearby glaciers or at ice margins. Numerous at the mammoth site, many had both shells attached, signifying that they perished without transport from another location. Other species common in clays at the site suggested a very cold water environment with fluctuating salinity.

Though soft-shell clams survive along a broad temperature range, they proved useful at the site for determining water depth at the time of Hairy-it's submergence. Their presence implied a depth of 20 feet or less in a bay or estuary environment.

Other invertebrate species supported one or both of the two indicator species mentioned. So it appeared that Hairy-it's remains flushed down a glacial river and sunk in the shallows just beyond the river's mouth.

34. Dealing With the Elephant in the Room

Hachaliah Bailey's first year of elephant ownership became a quick learning experience. The excitement of possessing such a wondrous creature soon collided with the demands of running a farm, managing a tavern, transporting cattle to market, and caring for his massive exotic pet. The African keeper knew the needs of an elephant, but the Bailey family had to supply at least one hundred pounds of hay, fifteen pounds of mixed produce, and eighteen to twenty-five gallons of water a day to keep the animal healthy.[1] Because elephants digest only about forty percent of their food, they produce prodigious amounts of manure, adding to the problem of confining Bet inside a barn for display. Yet despite these burdens, Hachaliah never doubted the importance of his purchase at the Bull's Head.

When word got around the neighborhood that the Baileys had an elephant, people flocked from miles around willing to pay whatever price was set to see this marvel. The gamble Hachaliah took to duplicate the success John Owen had with the Crowninshield elephant seemed to be paying off and at an investment of one tenth of what Owen paid.

Hachaliah purchased a moneymaker, but that was mostly because of her novelty, which he knew would soon wear off. To keep the cash coming in, Hachaliah had to travel with Bet, further infringing on time for his other responsibilities. He succeeded covering a circuit through Westchester, Putnam, and Dutchess counties because he knew many of the people in those areas through the livestock trade, social gatherings, and commerce at the Bull's Head Tavern, and could count on shelter and food.[2] But traveling farther afield required serious logistical planning. Part of that planning involved a trip south to acclimatize the elephant to North American winters. Whoever did travel south with Bet arrived in Lancaster, Pennsylvania in November then settled in Charleston, South Carolina, from January to February 1807.[3]

All the routes of travel had to be well thought out. Show stops needed the right accommodations – a barn for display, a bed, and meals. Posters and newspaper advertisements needed circulating. Time schedules needed locking in place so nighttime travel between venues seemed reasonable. Inns and taverns were considered the logical stopovers. Hachaliah planned to provide a cut for someone to travel with Bet and display her while the tavern and innkeepers profited from the thirsty crowd gathering to watch the elephant.

The winter after Bailey's initial tour with Bet must have been the time to work out details of Hachaliah's master plan. An improved road from Stephentown to Sing Sing known as the Croton Turnpike was being developed, and Bailey wanted to position himself to take full advantage of the toll road's access point and also the road coming in from Danbury and Peekskill. The key piece of property centered on Thomas Leggett's Inn at the north corner of the crossroad. Thanks in part to the success of Bet's local tour, Bailey had the funds to acquire the inn the following year, 1807, as well as over 350 acres across the intersection for $1,250 [4]. Two years later after the name change of Stephentown to Somers he controlled traffic to Sing Sing by way of a gate ordered by the town to "access the said road on Somers Town Plane, near the house of Hachaliah Bailey."[5] In another twenty years he would be elected secretary of the Croton Turnpike Authority.

On Tuesday, July 26, 1808, the following advertisement appeared on page 2 of the *New York Evening Post*.

NOW OR NEVER

A LIVE ELEPHANT to be seen at the house of Daniel Halsey, No. 56 Cedar – st. Those who would wish to gratify their curiosity, by seeing the wonderful works of nature, will do well to call immediately; perhaps the present generation may never have an opportunity of seeing an elephant again, as this is the only one in the United States, and perhaps the last time that he will be exhibited in this city; he is eight years old, and is upwards of seven feet high.

Admittance 25 cents Children half price

The price for seeing Bet must have increased from Bailey's initial fee of "a coin" for an entire family, if that coin was a quarter (traditionally called two bits). Also, the claim that the elephant "is the only one in the United States" may have been an advertising ploy, since there is evidence that the Crowninshield elephant was still alive then. Some reports do claim that an elephant was exhibited in southern Canada at that time, which, if true, most likely was the Crowninshield elephant.

Just over two weeks later Bailey made a hand written lease agreement that now resides in the Somers Historical Society Collection.

Article of agreement between Hachaliah Bailey of the first part and Andrew Brunn and Benjamin Lent of the second part. This Brunn and Lent agree to pay this Bailey twelve hundred dollars each for the equal two thirds of the elephant for one year from the first day of the month. Bailey on his part furnishes one third of the expenses and Brunn and Lent the other two thirds.

August 13th 1808

That December 17th a similar ad to that of July 26 with minor modifications ran in a New York newspaper of the showing of the elephant at Wm. Satterwhite's establishment.[6] Apparently, suitable accommodations had been made to keep Bet up north, at least for that month.

35. The Dynamic Homeland of Hairy-it

Through many decades of challenging work by geologists we now know that approximately 25,000 years ago the Laurentide Ice Sheet reached its southern Maine limit on the continental shelf 250 miles southeast of Scarborough. By 21,000 years ago the ice front began melting back faster than it could advance. It reached the present coastline roughly 3,000 years later, an eye-blink in geologic time. But if you had been at that point back then, you would have been in a boat dodging icebergs.

As the mile and a half thick ice buckling the continental crust thinned, land rebound lagged behind the pace of ice-melt, allowing ocean water to chase the glacier's front miles inland from our present shore.

Imagine a huge flatbed truck hauling a wide conveyer belt fed by a hopper loaded with a mix of mostly sand, some gravel, and a few rocks. The conveyer hangs from the back of the truck where it dumps its load onto a flat road flooded with about a foot of standing water. While the conveyer is running, the truck moves slowly along. This conforms in a very simple way to the retreat of the glacier as it dumped its load of debris into the invading ocean. Periodically, the truck stops to receive more mix from a front-end loader so that the hopper doesn't run empty. The conveyer belt keeps in motion at all times. By mid-morning the truck driver stops to take a coffee break in his cab, then a lunch break at noon. Later he takes another break before shutting everything down for a long holiday weekend. The buildup of debris dumped during each break simulates periods when the glacier's advance and melt-back were in dynamic equilibrium.

Just prior to the driver returning to work, we glide slowly above the flooded road in a glass bottom boat that leaves no wakes as we follow the truck's course during the last day of work. At first we see a mostly even surface of fine sand with intermittent peaks caused by buried gravel or rock. Where the truck stopped at various lengths of time, linear ridges the length of the conveyer belt's width and with depths corresponding to the length of time during each stop are also covered with fine sand.

A wind sweeps across the flooded road, helping to dry the pavement. In so doing, it constantly disturbs the fine sediments covering the bottom. The finest deposits are washed into the lowest portion of the roadbed when the pavement finally dries.

In this crude model the wind winnowing represents a complex of meltwater turbulence, land rebound, and weather dynamics during the glacier's retreat over Maine.

One of the consequences of the glacier constantly plucking stone and grinding it against bedrock when the ice advanced were grooves left in the rock face still visible today in some locations, allowing geologists to document the direction of ice flow. The other was a copious amount of fine debris produced by this process.

When the glacier front melted back, its base exposed a jumble of rocks, gravel, and sand called glacial till. At the same time meltwater flushed clouds of grindings into the encroaching sea. The finest of these particles slowly settled to form clay beds burying the till. Icebergs calving off the glacial front dropped a diverse assortment of stones that sunk into the clay cushion. During cool periods when the ice advance kept pace with meltback, rock and gravel piled up at the base of the underwater ice front, sometimes in enormous linear rows bordering the glacier. That too received a clay coating, as did the submerged

Scarborough site, where the rain of fine grindings settled as mud that transformed into deep clay beds.

While the sea lapped at the glacier's retreating front and icebergs calved away, the ocean grew shallower over Scarborough from crustal rebound. Eventually the bases of storm waves churned up mud from the bottom that later resettled into interrupted layers of fine sand covered by a finer mud. This wave sorting continued more frequently as the sea shallowed, and in one of those sand lenses, the remains of Hairy-it were sealed.

36. A Blast From the Tropic Past Affects an Elephant Lass

Hachalilah Bailey's financial success from leasing out a portion of his elephant allowed him to run his farm and pursue other ventures while counting on a steady income from Bet. He could not foresee that an incident 10,000 miles from Somers, New York, would contribute to the demise of this cash flow.

At the turn of the 19th century the island of Sumbawa already had a long history of fame. Situated midway along the Malay Archipelago[1] it possessed a climate halfway between the extremes of tropical jungle at Sumatra, farthest west on the island chain, and the seasonally dry forests on Timor, far to the east. Tree studded savannahs, interspersed with cultivated strips, carpeted the lowlands like quilt patches stitched together by threads of a dark ancient forest spilling down from the peaks. Its longstanding pride centered on horse breeding which had been celebrated since well before the 16th century. The rainy season, spanning the months from November to the first of April, spawned luxurious emerald pastures across the savannahs, but the grazing paradise shriveled as the dry season advanced leaving only the fittest horses to graze the lush grass again. Over the years this cycle of selection created horses "famed for their stamina and endurance."[2]

Sumbawa was also known for its sappan wood indigenous to the island, which produced a stable red dye coveted by many nations including those in Europe.

When the Dutch controlled trade in the region during the 18th century, a coffee plantation took root on the west slope of the largest land feature on the island known as the volcano Tambora. At an estimated height of 14,100 feet, the mountain functioned as the dominant navigational feature for ships sailing along the island chain. Long thought to be extinct, Tambora began showing signs of life in 1812 when the earth rumbled and a dark cloud appeared at its summit. Two years later John Crawfurd, a British diplomat and naturalist, sailed by the volcano and noted, "At a distance, the clouds of ashes which it threw out blackened one side of the horizon in such a manner as to convey the appearance of a threatening tropical squall. ... As we approached, the real nature of the phenomenon became apparent, and ashes even fell on the deck."[3]

The mountain quieted down but again came to life on the evening of April 5, 1815 like a monstrous hammer striking the earth. Flames shot skyward from the summit and lava began spewing to percussive, ear-splitting blasts that to witnesses 900 miles away sounded like cannon fire. According to Sir Thomas Raffles a British colonial agent stationed on Java at the time, "... as the eruption continued, the sound appeared to be so close, that in each district it seemed near at hand...From the 6th, the sun became obscured: it had every appearance of being enveloped in fog...explosions continued occasionally, but less violently, and less frequently than at first."[4]

The worst seemed over, and people from the village of Sanggar near the base of Mount Tambora rushed to clean away ashes from their crops before they smothered. As the day ended on April 10th, they left the fields relieved that their plants had survived and unaware that the gates of Hell would soon open around them.

While people settled into their homes around the time of the evening meal, an earsplitting jolt rocked their world. "On the island of Sumatra, over a thousand miles west of Tambora, local chieftains heard the explosion."[5]

The Raja of Sanggar rushed from his home and looked to the summit of Tambora. Three mammoth tongues of fire roared from the summit's crater and coalesced into an inferno that seemed to burn away the sky. He and his family fled on their fastest horses. Forty miles eastward from the volcano in the town of Bima a resident claimed that the blast was like, "the report of a heavy mortar close to his ear."[6]

> In a short time the whole mountain next (to Sanggar) appeared like a body of liquid fire extending itself in every direc-tion...darkness caused by the quantity of falling matter obscured it at about eight, P.M. Stones at this time fell very thick at (Sang-gar); some of them as large as two fists...Between nine and ten P.M. ashes began to fall...[7]

The super-heated air above the earth's rupture rose so quickly that a vacu-um formed at the surface drawing in a thunderous rush of air to fill the void. A violent whirlwind ensued...tearing up by the roots the largest trees, and carrying them into the air, together with men, houses, cattle, and whatever else came within its influence.[8]

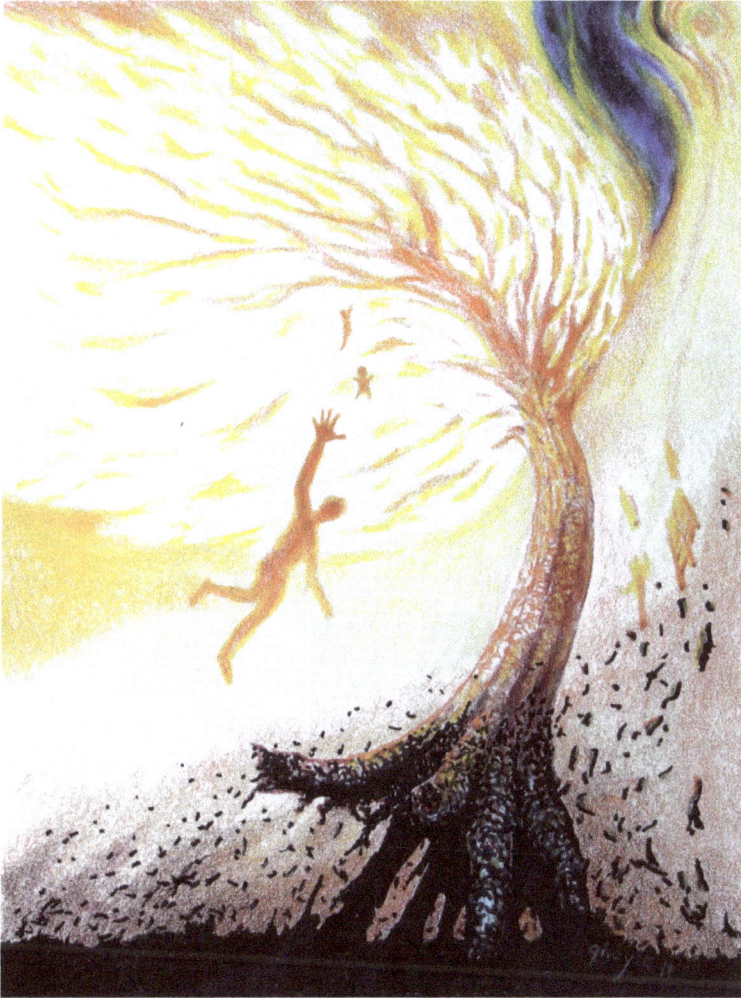

Tambora's wrath. (Author's illustration)

As ejecta roared into the stratosphere, pressure dropped in the magma chamber causing the earth to buckle and subside. The shock created a twelve-foot tsunami that swamped the coast and collided with the volcano's pyroclastic flow. The explosive power of their embrace generated flashes of superheated steam primed to fatally scald any remaining survivors at Sanggar.

In the dark hours of morning on July 11th about 250 miles to the northeast of Tambora, the cruiser *Benares*, anchored at Makassar in the Celebes (now Sulawesi), shook from repeated blasts of what the Commander feared might be nearby cannon fire from marauding pirates. Men sent to the mastheads saw no flashes. At dawn the *Benares* weighed anchor and headed south.

The following morning a disturbing darkness began to hang above the vessel and deepen as the minutes passed. By 8:00AM intense blackness blotted the south and west and raced across the heavens like an unfurling sail. Then showers of ash began falling on the deck at such a rate that it frightened the Commander and crew. And still the darkness deepened.

By noon the last vestige of light on the eastern horizon died, and an inky blackness ruled the rest of the day. The Commander recalled, "I never saw any thing to equal it in the darkest night; it was impossible to see your hand when held up close to your eyes."[9]

Next morning diffuse sunlight seeped through the volcanic fog, revealing a ghostly deck covered in a foot of volcanic ash.

The crew dumped what the Commander estimated to be several tons overboard, while a light rain of ash continued falling throughout the day. Around noon the sun could be seen through the murky haze that lingered for nearly a week.

At the same time that the *Benares* sailed in midday blackness, a Malay pilot glided his proa through the gloom. As he approached Sumbawa, the sky brightened enough to see Tambora five miles away. Fire raged on the lower slopes while thick pendulous clouds hid the summit. Curious and in need of water, the pilot went ashore and found a landscape covered in three feet of volcanic ash. Large proas thrown inland by the tsunami lay helter-skelter among the dead inhabitants.

> On leaving Sumbawa, he experienced a strong
> current to the westward, and fell in with great
> quantities of cinders floating on the sea, through
> which he with difficulty forced his way; he was
> surrounded by them the whole of the night of
> the 12th, and says they formed a mass of two feet
> thick, and several miles in extent.[10]

The rafts of volcanic cinders that the pilot encountered persisted for years after the event and spread as far as the Indian Ocean.

The dominant ash fall and pyroclastic flows spilled west and north of the volcano's caldera, where the thriving minor kingdom of Tambora had existed just days before. Researchers found evidence of this domain in 2004 under ten feet of debris. Within the charred ruins of a house, they discovered the flash carbonized bodies of a man and woman positioned as though ready to share a meal. Lead researcher Haraldur Sigurdsson called the site the "Pompeii of the East."[11]

Just a few days after the April 10th eruption the ash cloud spread to the size of Australia, chilling the tropical air in some places to temperate level. One to three feet of ash dispersed over the western half of the island chain. Then the famine began. From the direct blast to the end of Tambora's induced famine, upwards of 100,000 people died and untold numbers of fish and wildlife. The eruption is considered the greatest volcanic event in the last 10,000 years. The mountain went from a height of roughly 14,000 feet to a little over 9,000 feet, ejecting 9.8 cubic miles of volcanic debris and leaving a crater 3.7 to 4.3 miles across and 2,000 to 2,300 feet deep. The ash cloud reached an estimated height of 144,000 feet where the finer particles and sulfurous gasses dispersed and affected climate worldwide for years to come, most significantly in 1816, when elephant Bet journeyed into the District of Maine.[12]

37. New Eyes on America's Ice Age Prize

By 13,000 years ago the glacier front had been retreating for thousands of years, resulting in an acceleration of land rebound. Mammals new to the Americas had migrated across the land bridge and established themselves on land south of the ice sheet. One species, an apex predator, thrived in the new hunting grounds; a human population came into the land in several waves over many years. The exact beginning of the migration is still debated and so is the route of the arrivals.

When I began to investigate the Scarborough tusk, the exhibit hall where I worked, "12,000 Years in Maine," celebrated the standard theory of that day, which suggested that a land corridor opened from Alaska to lower North America about that time period, allowing humans to migrate south after crossing the land bridge from Siberia. These people appeared to have entered the Americas as skilled big game hunters known as the Clovis people, so named from their fluted stone spear points being first found in Clovis, New Mexico. Proponents of the Clovis First Theory claimed these hunters swept across the Americas driving megafauna naïve to human predation to extinction.

A paleo-American hunt. (Author's illustration)

But more recently that theory has been effectively challenged.

During the last Ice Age the dominant Laurentide Ice Sheet, originating in Hudson Bay, merged with one coming off the western mountain chain known as the Cordilleran Ice Sheet. When their glacier fronts retreated, an unzipping of the ice sheets occurred from south to north leaving a glacial till gap between them, which was the ice-free corridor scientists envisioned as the path taken by humans first entering the Americas. But the timing of the opening no longer supports the Clovis First Theory. Recent studies suggest that the corridor remained unsuitable for human migration until about 11,500 years ago: long after the arrival of the first people.

In 1977 anthropologist Tom Dillehay excavated a site found two years earlier by a veterinary student near a small tributary of the Maullin River, thirty-six miles from the Pacific Ocean in Monte Verde, southern Chile. The site stayed exceptionally well preserved because the rising water table created a bog over the encampment. In 1982 radiocarbon dating of bone and charcoal from the Monte Verde site gave an average age of 14,800 years, over a thousand years older than the dates for the earliest Clovis hunters. At a conference on the settling of the Americas at the University of Maine in 1989 Dillehay gave a presentation on his Monte Verde findings to a stunned audience of professional researchers, but the dating at 14,800 years ago remained unaccepted until 1997 when a group of prominent archaeologists examined the site.[1]

About the same time that Dillehay gave his talk at the University of Maine, archaeologist James Adovasio at the Meadowcroft Rock Shelter site near Pittsburg, Pennsylvania, waited for a third party to retest samples for radiocarbon dating that he had had done but were rejected in the past by the broader archaeological community. The concern centered on the possibility that water seepage from coal beds skewed the original results. The particular samples came from the base of what had been a continuous seasonal settlement up until the mid-1800s. The samples tested at 16,000 to 18,500 years old. The more recent testing done by a third party revealed that no contamination had occurred and under more stringent analysis the dates held.[2]

Other sites also showed ages older than the traditionally accepted date. The Buttermilk Creek Complex in Texas, excavated by Michael Waters in 2006, included over 15,500 artifacts dated in a range of 13,200 to 15,500 years ago. Waters also reanalyzed a rib from the Manis Mastodon site in Washington State and found that a projectile point reported in the rib was a bone weapon. Waters reported the original radiocarbon date valid. His own sample dated 13,800 years ago.[3]

Then in 2021 the journal Science published a paper by geoscientist Matthew Bennett and colleagues on partially buried human trackways at White Sands National Park in New Mexico. A portion of their abstract reads "...*in situ* human footprints are stratigraphically constrained and bracketed by seed layers that yield calibrated radiocarbon ages between ~23 and 21 thousand years ago." If these dates hold up to future scrutiny, they will deal a major blow to The Clovis First Theory.

Sites with more skeptical data continue to challenge the traditional view, but none provide such a shocking claim as a site my friend, Dan Fisher, worked on in San Diego County, California, known as the Cerutti Mastodon Site. Originally considered a paleontological site constituting the remains of extinct dire wolf, horse, camel, mammoth, and ground sloth, one assemblage of juvenile mastodon bones looked out of context from a purely paleontological viewpoint. The bones appeared deliberately smashed to extract marrow, and near them scientists found what looked like hammer stones and anvils. Micro-abrasions on the stones and skeletal elements matched, implying human activity. More than twenty years after the site's discovery in 1992 the mastodon bones were dated using a state-of-the-art radiometric thorium-230/uranium dating technique rather than radiocarbon dating which is only accurate to an age of about 50,000 years. The result stunned everyone: 130,000 years old. No site in the Americas comes close to that date. If it holds up to future third party investigations, the Cerutti site will predate the accepted time of human migration into northern Asia and imply that Neanderthals or possibly a lesser-known species called Denisovans came into the Americas. Whatever species processed bone at the Cerutti site, the only available route for them to arrive there at that time would have been by water, dodging the giant ice flows.[4]

Coastal camp. (Author's illustration)

A water route has been suggested for later arrivals prior to the opening of the ice-free corridor. Jon Erlandson and his colleagues in 2007 proposed that people traveled from northern Asia into the Americas on a "kelp highway" that stretches to Baja, California then flourishes again along the Andean coastline. Kelp forests are highly productive ecosystems rich in aquatic life from invertebrates and fish to sea birds and mammals, such as seals and sea otters. In the past the extinct thirty-foot Stellar sea cows became easy targets for hunters in these northern Pacific underwater jungles. Kelp beds also dampened the force of the open sea and flourished on the nitrogen rich meltwater that flowed off the ice sheet.[5]

Whatever way paleo-Americans got here, by the time the corridor from Alaska to the lower states opened for migration, the land south of the ice sheets had been occupied for untold generations[6], and the impact of humans on the landscape was accelerating a major extinction event.

151

38. It's a King Solomon Thing

With the completion of the Croton Turnpike Road, travel to and from Sing Sing increased substantially. The tollgate erected next to the inn Hachaliah Bailey bought from Leggett buzzed with activity as did the inn itself. Pressure to accommodate the increased number of guests put strains on everyone working at the crossroads. It may have contributed to Hachaliah building a home in 1810 some distance from the turmoil, on his newly acquired 350 acres. The house, a two-family structure, was erected on a hillside that would later be known as Elephant Ridge.[1] The spacious barn may have periodically become the "hideout" of Bet when she came off tour, though that year she was located in Savannah, Georgia, by October 25,[2] and there is evidence she overwintered in Albany, New York, at Wetmore's Inn in 1813 and 1814.[3]

On October 11, 1811, Benjamin H. Latrobe, an architect who worked on the United States Capitol, spent an overnight at Scholl's Tavern in Clarksburg, Maryland, where he encountered an elephant on display. In his sketchbook he jotted down details about the curious-looking creature.

The elephant was a ten-year-old female, eight-and-a-half feet tall. Latrobe noted that her tusks had been "broken off." Whether they broke by accident or were cut back was not mentioned. However, he did state that they were growing and had lengthened an inch over the last three months. He also said the elephant could crack a whip, uncork a bottle and draw its contents into her trunk before spraying it into her mouth.

Along with this description, Latrobe included two watercolor illustrations. One was a double profile of the elephant showing her in a passive and an active mood. The other showed the elephant being led away from the tavern at night along with a small group of people.

Today a public park sits where the old tavern once stood. In that park a plaque displays photographic reproductions of Latrobe's images of the elephant with his written description alongside them. In the lower right-hand corner a contemporary write-up claims the elephant was the one Jacob Crowninshield brought to America in 1796. But that is in error, and here is why.

Crowninshield's elephant would have been 16 or 17 years old in the fall of 1811. Bet was 10. Furthermore, Crowninshield's elephant was Asian (specifically from India). As noted by Rev. William Bentley in Chapter 29, it was a female with tusks protruding only slightly "beyond the flesh." Noticeable tusks are rare in female Asian elephants. On the other hand, the presence of tusks, the severe slope of the head, and large ears are typical of African elephants of both sexes. Since only two elephants were present in North America in 1811: one from India and one from Africa, and since they were both females, Latrobe must have painted illustrations of elephant Bet. Readers can verify this themselves by doing a Google search for elephants then Googling Dowden's Ordinary Park to see the plaque emblazoned with photographs from Latrobe's sketchbook.[4] Latrobe's sketches constitute the only verifiable illustrations of this historic animal. His Sketchbook is now archived at the Maryland Historical Society.

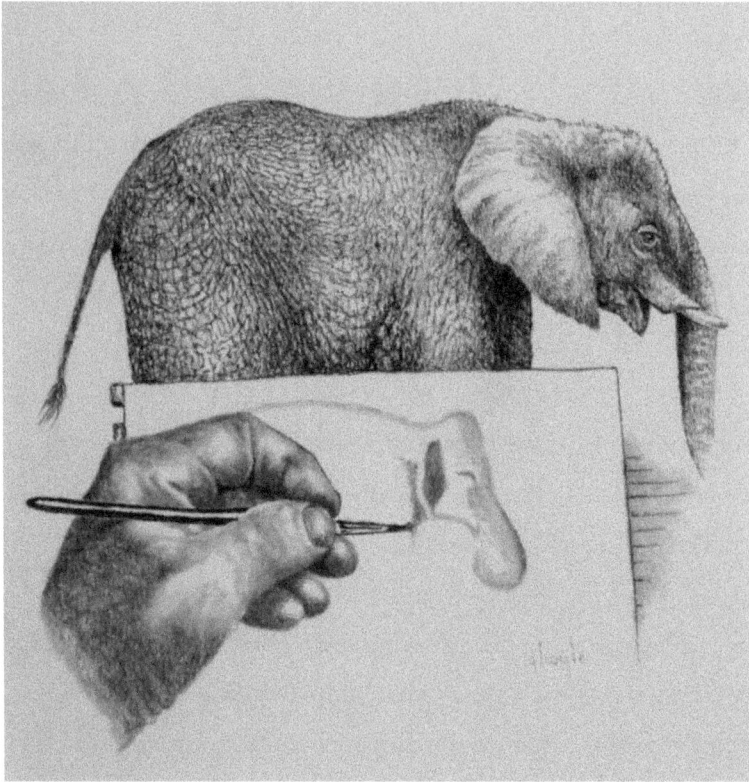

Author's interpretation of Benjamin Latrobe sketching elephant Bet.

While Benjamin Latrobe worked on his sketches of Bet, fifteen-year-old Nathan Howes balanced on a tightrope trying to perfect his technique. His family ran a general store in Brewster, New York, and had a farm just outside of town. He and his five brothers and six sisters learned to work the farm and store at an early age. By the time they reached their teens it had become second nature to them. But Nathan dreamed beyond the farm and store. He loved to be the showman of the family and became fascinated by tricks he had seen practiced by traveling performers. Cuts and bruises didn't discourage him from being a daredevil and he gravitated to funambulism especially when he found out that tightrope walking had a long noble history and such a fun-sounding formal name. He did, however, take his practice sessions seriously.

After his fifteenth birthday on April 22, 1811, Nathan met an itinerant funambulist and became his assistant at the price of one dollar a day. During that year he perfected his art and learned hat-spinning tricks. His whole family soon took an interest in circus arts and some of them joined Nathan and his friends in forming The Nathan Howes Roadshow, at which point Nathan's funambulist mentor became part of the show.[5]

The following spring as plans crystalized over a schedule of local venues for the nascent show, rumors of war with Britain overshadowed everything. Then on June 18, hostilities led to President Madison signing a declaration of war. Sixteen-year-old Nathan Howes soon became one of the able-bodied men responding to the call. Whatever role Nathan played during his two and a half years of military service, it did not impact his love of the circus. Soon after the United States Senate ratified a peace treaty with Britain on February 17, 1815, he began to plan for the resurrection of his Roadshow. He returned to his training haunt, an old ballroom at Sodom only a few miles from his family's farm, to practice fervently on the wire rope he had strung there in the autumn of 1811.[6] He also trained his dog to do tricks and sought out other animal acts.

By the time Nathan traveled again with his Roadshow, it had taken on an air of professionalism that brought praises across the counties where the troupe traveled. Nathan's act dominated the show. In time he would be considered the finest tightrope walker of his day. But as The Nathan Howes Roadshow ended its tour that year, the ambitious nineteen year old hoped his profits would add a major highlight to next year's show, which would compel him to extend his travels throughout the Northeast.

Hachaliah Bailey's success leasing out Bet had become legendary, but those lease agreements had been restricted to Hachaliah's close relatives and friends.

Nathan hoped his own success with the roadshow would persuade Bailey to take a chance on him. Living about a dozen miles from the Howes farm, Bailey knew the family and sometimes frequented their general store. He knew of Nathan's skill as a funambulist and may have watched him practice at the old Sodom Dance Hall or attended a Nathan Howes Roadshow performance. So he was not surprised when Nathan contacted him and offered a plan to tour Bet north through areas where no elephant had been. Hachaliah liked the idea of combining the elephant exhibition with a circus, and together along with suggestions from Bet's African handler, they worked out the details of a northeastward tour all the way into the District of Maine.

By early April a legal agreement had been signed between the parties giving half ownership to each during the 1816 tour. Taking Hachaliah's advice Howes left at night with his Roadshow caravan of friends and relatives[7] along with the addition of an elephant, a wagon carrying a breakdown shed to hide Bet when necessary, and at least one hay wagon.

As The Nathan Howes Roadshow and Bet disappeared into the seasonably warm night, Hachaliah turned his attention to his many ventures involving the transportation of people and animals through Somers to New York City. But after a few weeks of not hearing from Howes, he became concerned as to why he had not received his remittance from the Roadshow venues. He wrote to Nathan but got no reply. He then took the stage over a hundred and seventy-five miles to Boston. There he learned that large crowds of people had seen the elephant. A few days later on Saturday the 27th he caught up with the Road-show in the whaling town of New Bedford at the mouth of the Acushnet River, where he witnessed hundreds of people visiting Bet's exhibition.

By the time he found Nathan his patience had grown thin as a razor and his temper just as sharp. With a fight for composure he demanded an immediate settlement. Howes retorted that he had no time for it and would settle up when he returned in autumn. Bailey became wary, suspecting that Nathan might not come through with any of his payment. After further haggling, he offered to sell his interest in Bet to Howes. By now Nathan was aware of the maintenance demands of the animal and replied that he had elephant stock enough. Bailey fought to contain himself as he proposed to buy back the lease, but Nathan expressed his satisfaction with the present arrangement. With that remark Bailey became furious, shouting that as long as he had any interest in the elephant, he would not allow it to travel any farther. Howes stood defiant and insisted that Hachaliah could not prevent it because their legal agreement stipulated that he had charge of the elephant until they were to settle up in autumn. Bailey fired back that it also stipulated that he receive half the profits as they accumulate.

Howes stood his ground and bellowed, "There are no profits!"

At that Bailey's voice lowered to a growl as he vowed to make sure the young man took Bet no farther. Then he walked away.

The following morning as the Roadshow crew packed up, Nathan went to the stable to get Bet ready for the journey. As he stepped into the dark interior, the light from an oil lamp drew his attention. Next to the elephant stood Bailey with a loaded gun. When Nathan stepped closer Hachaliah raised the barrel. Howes froze. Was the gun meant for him?

Bailey gave the twenty-year-old a hard look and answered, "No."

In a steady low voice he continued, "I intend to only do what is lawful."

Swinging around he pulled the gun to his shoulder and aimed at the elephant's head.

Nathan cried out, "Half that elephant's mine."

"I'm only going to shoot my half," Bailey replied.

Bailey's threat. (Author's illustration)

Fearing it was no bluff, Nathan begged Hachaliah not to fire and promised to settle immediately.

The two men parted with terms that pleased Bailey, and the Roadshow continued on its way with Bet.[8]

This paraphrase of a published account decades after the event gives the impression that Nathan Howes changed his pattern of travel to daylight hours. However, with the concern of frightening horses and providing the public with a free peek at Bet, it is doubtful this was a normal occurrence. With New

Bedford being a shipping as well as whaling port and Sunday blue laws in effect that severely restricted roadshow entertainment, it's more likely Nathan took advantage of the downtime by getting the elephant ready to travel on a ship to the next venue. It justifies his reluctance to settle with Bailey and why he was allowed to continue his tour with Bet. Ship passage could not have been cheap for wagons, crew, and an elephant.

39. A Fragile Steppe for Hairy-it

During the last Ice Age, vast arid grasslands persisted from Europe to Siberia and across a wide land bridge into parts of Alaska. Deeply rooted grass species intermingled with low, herbaceous plants. These herbaceous plants known as forbs were specially adapted to the windswept, bone-chilling, and often dusty environment that dominated the region. Many different grazing animals and predators roamed the area, but this environment, unique to that time period, is named after one species symbolic to the Ice Age. Scientists call this vanished ecosystem the mammoth steppe.

In North America the mammoth steppe also spanned the continent just south of the ice sheet over what now constitutes the northern contiguous states. The woolly mammoth was so well adapted to this environment that it unlikely migrated any great distance south during the colder seasons. As mentioned earlier, abrasions found on the lower parts of many tusks have been interpreted as resulting from mammoths breaking through ice on frozen waterways and sweeping snow from buried plants.

In summer, even when advancing, the ice sheet shed water from its glacial front. Rivulets merged into braided streams that coursed into meltwater rivers, but water also seeped below the mammoth steppes nourishing the deep rooted vegetation with moisture rich in nitrogen that had been locked for hundreds of centuries in the glacial ice.

The seasonal blooming of the mammoth steppe created a lush paradise for grazing animals. Recent research on the remains of this ancient environment shows that the mammoth steppe was as productive as the present day African savanna where huge herds of herbivores congregate after the rainy season.

Similar migrations must have poured into the mammoth steppe to graze and calve alongside the shaggy tuskers.

Horses, camels, bison, and mammoths – just some of the mammoth steppe grazers. (Author's illustration)

Dark clouds of ancient bison and giant camels mottled the grassland along with reindeer, horses, muskoxen and perhaps pronghorns, and giant ground sloths. Just as on the present day African savanna, the abundance of prey species attracted an array of predators, from wolves and lions more massive than today's species to sleek cheetahs unrelated to, but looking much like, the present day African cat. The giant short-faced bear, called by some an overgrown grizzly on stilts, roamed the land on its long legs as an omnivore or opportunistic predator. These unique grasslands existed as a stable ecosystem for thousands of years, but as the ice sheet collapsed so did the mammoth steppe.

Ever since its discovery by Dale Guthrie at the University of Alaska Fairbanks in 1980,[1] the exact structure of the mammoth steppe has been debated.

Today's tundra is a far wetter environment than the Ice Age steppes even though the annual precipitation is low in the region. Due to extensive subsoil permafrost the surface soil does not drain well in the summer. Fibrous plants rich in toxic alkaloids make up part of the ecosystem along with shallow rooted grass species and succulent forbs adapted to surviving a cold, windy environment with seasonal extremes of light and darkness. Many mosses and lichens carpet the landscape. Because of the frozen subsoil, roots cannot penetrate deep enough to sustain trees in this Arctic region. In fact the word tundra derives from a Finnish name "tunturia" meaning treeless or barren plain.

The specially adapted tundra plants limit the types of animals that can graze on them and thrive in such a hostile environment. Long winters tax the hibernators, but long summers with nearly continuous sun stimulates abundant plant growth that encourages migrations of birds and certain mammals from the south, journeying into the area to fatten for winter, give birth and raise their young for the exodus south.

Some animals of the mammoth steppe adapted to a changing world and now inhabit various parts of the Arctic tundra. Among them are the caribou, other reindeer, and the muskox. Whether these species shifted their diets with the loss of the mammoth steppe or were niche feeders on the steppes is unknown presently.

In the 1990s when a few scientists seriously considered the possibility of resurrecting a mammoth, many others ridiculed the idea because even if it were successful, the mammoth steppe remained an extinct environment that couldn't be revived. That criticism launched several research projects to determine precisely what constituted the mammoth steppe and to search out regions that might harbor remains of it. In describing the makeup of the mammoth steppe, Dale Guthrie noted that the record of pollen (which is resistant to decay) and plant fossils imply an open landscape of low growing herbaceous plants dominated by grasses, sedges, forbs (such as yarrow, mums, and tansies), and sages.[2]

A later study led by Danish scientist Eske Willerslev of a fifty thousand year period of plant growth in the Arctic partially reinforced Guthrie's appraisal of the steppe's composition, except the study revealed that grasses played a subservient role to the more nutrient rich forbs on the landscape. Samples from the first twenty-five thousand years showed a wild diversity of these plants until the variety declined during the Last Glacial Maximum. Nevertheless, forbs remained dominant on the mammoth steppe to the end of the Ice Age.[3] Because grasses produce a greater preponderance of pollen than forbs, it skewed previous studies of the steppe's plant composition based on pollen count.

Other studies found that the makeup of the plant community at that time consisted of some extinct species and those that today no longer associate with other flora of that period, implying the mammoth steppe had a dependent relationship with the glacial front. Yet fragments of that plant community seem to have survived in the northern part of central Asia and isolated parts of North America.

Aware of these rare environments, Russian ecologist Sergey Zimov had the idea that it might be possible to coax back a form of the mammoth steppe by increasing the density of grazing animals on the modern tundra. Warm events in the distant past, when the glacier fronts retreated significantly, resulted also in the mammoth steppe's retreat, yet throughout these numerous interstadial events the mammoth steppe survived. Zimov reasoned that the loss of mega-fauna at the close of the Ice Age might have been the primary cause of the mammoth steppe's collapse. After all, megafauna on the present day African savanna do limit tree growth by uprooting and trampling vegetation during high-density migrations. To test his theory, he developed an experiment he called Pleistocene Park on a 50 square mile tract of land in Siberia, where he released herbivores in higher densities than normally exist today on the tundra, hoping that trampled mosses and lichens would give way to grasses and forbs. Mechanical equipment uprooted the low-growing trees, mimicking the behavior of elephants and what is thought to have been a similar behavior in mammoths.

Zimov began his grand experiment in 1996. Eleven years later a reporter visiting the park for *The Atlantic* magazine wrote:

> [The park] has broken out of its original fences, eating its
> way into the surrounding scrublands and small forests.[4]

It appears that Zimov is making progress, but Dale Guthrie cautions:

> Though large mammals themselves may have had important effects
> on the steppe vegetation, they probably were not the key in produc-
> ing and maintaining it.[5]

Zimov responded in the journal *Paleontology in Ecology and Conservation.*

> Herbivores enhanced biocyclicity, trampled moss and shrubs
> and maintained pastures. Therefore this ecosystem was only
> partially dependent on climate.[6]

40. An Elephant's Chilly Walk

After Nathan Howes settled his debt at New Bedford, Hachaliah Bailey allowed him to continue his tour with Bet. Howes took his Roadshow north on a coastal schooner. As it sailed that Sunday, a dry fog settled over the water. The acrid odor gave Nathan the sense that a forest might be burning somewhere toward the north.

The Nathan Howes Roadshow arrived in Salem Harbor early on Monday. The ever-inquisitive Reverend William Bentley came down to meet the boat and got permission to examine Bet prior to her public exhibition. He jotted in his diary, "I went in the morning when I might examine him without any of the tricks he has been taught to play."[1]

As in describing the Crowninshield elephant, the Reverend used the pronoun "he," though clearly seeing the proof of her sex. In many ways she resembled the Crowninshield elephant, but the most striking difference centered on the length of her tusks, which protruded much farther from the gum line.

As Benjamin Latrobe and the Reverend noted, Bet had become more than a curiosity; she played the role of performer and seemed to relish it, unlike what Bentley penned about the Crowninshield elephant when he saw it: "The keeper repeatedly mounted him, but he persisted in shaking him off." [2]

By the latter half of May, Howes' troupe had entered the District of Maine. Fliers went out ahead of them perhaps distributed by a member of Nathan's roadshow. Newspapers were notified a week in advance and through previous arrangements suitable locations were set to show the elephant. Traveling by night The Nathan Howes Roadshow arrived at its destination with all accommodations ready, usually at an eager keeper's tavern with a spacious barn.

On May 23rd and 24th Howes showed Bet at Major Jefferds' Tavern in the shore town of Kennebunk. He then went a few miles up the coast to Biddeford. A woman from the neighboring town of Saco later recalled, "In April of that year the snow was five feet deep, and the ground was frozen solid. In May the ground was partially frozen, and in no condition for seeding."[3]

It was not just a local aberration. A deep chill seeped southward. In Richmond, Virginia, a local newspaper, the *Eastern Argus* reported, "On Thursday morning (May 29th) we had a frost in the city."[4] A similar wave of cold settled on the shore in New York Harbor. No one at that time could fathom that Tambora's eruption dialed back the season. It seemed inconceivable that dust launched into the stratosphere on the other side of the world could have any effect on weather at a global scale. Spring just seemed to be coming later than usual.

By the time the Roadshow reached Portland, the *Eastern Argus* had printed the following advertisement headed with an engraving of an elephant with tusks:

NOW OR NEVER!!

A FEMALE

ELEPHANT

TO be seen at TIMOTHY BOSTON'S COLUMBIAN TAVERN, In Portland, from the 29th inst. Until the 4th June.

THE ELEPHANT is not only the largest and most sagacious animal in the world, but from the peculiar manner in which it takes its food and drink of every kind, with its trunk, is acknowledged to be the greatest natural curiosity ever offered to the public. The one now offered to the view of the curious is a female. She will draw the cork from a bottle, and with her trunk will manage it in such a manner as to drink its contents. She is fifteen years old, and measures upwards of twenty feet from the end of her trunk to that of her tail; thirteen feet round the body; upwards of eight feet high, and weighs more than six thousand pounds. Perhaps the present generation may never have the opportunity of seeing an ELEPHANT again, as this is the only one in America, and this perhaps is the last visit to this place.

Admittance 25 cents – Children half price. Hours of Exhibition from NINE in the morning until SEVEN in the evening.

The claim that Bet was the only elephant in America at that time may be true. In his diary entry of January 9, 1816, Reverend William Bentley penned, "visited the Elephant kept in Boston." That animal was undoubtedly the Crowninshield elephant; Bet did not arrive in Boston until sometime in mid-April. Owen, who purchased the Crowninshield elephant, traditionally took it south in the winter to avoid the extreme cold. Why he settled in Boston with the elephant for winter is perplexing. Did Owen or his elephant suffer from illness? After Bentley's date, no reliable information about this animal could be found.

While Howes exhibited Bet in Portland, William M. Brooks at the southeast border of Maine wrote in his journal, "Lubec, May 29th...snow this morning and part of last night...May 31. –Saw ice in the road at Lubec as thick as a cent, when the sun was three hours high." He also recorded snowstorms on the second and sixth of June.[5]

At that time, unseasonably cold weather and violent rainstorms swept across Europe. Eighteen-year-old Mary Wollstonecraft Godwin wrote to a British friend describing her view of it from the Villa Diodati at Lake Geneva: "June 1. The thunderstorms that visit us are grander and more terrific than I have ever seen before."[6]

She had traveled from Britain with her recently divorced lover, the poet Percy Shelley, to enjoy a summer in the romantic grandeur of the Swiss countryside, but the bucolic landscape slumped into gloom during the months that dominate the summer season. The famous/infamous British poet, Lord Byron, had invited them to join him for the summer, so they rented a cottage on the water not far from the Villa Diodati where Byron vacationed with his young personal physician, John Polidori. During that time the weather inspired them to gather and read aloud some German ghost stories, an experience that prompted Lord Byron to challenge them to each write a haunting fiction of their own. Their creative energies birthed two classic tales of horror: Polidori's dashing and seductive *Vampyre* and Mary Shelley's plunge into the obsession of reanimating the dead and its resulting consequences in *Frankenstein: or, The Modern Prometheus*. Though the genius of these stories cannot be credited to the weather, it set a somber and foreboding fertile ground to explore the Gothic.

41. Killing Time on the Mammoth Steppe

It seems obvious that a link exists between the Ice Sheet and the mammoth steppe; it's easy to assume that climate change was the sole culprit in the steppe's demise. But just like Waldo Pray's challenge that the bones of Old Bet lay in that Scarborough gulley, Sergey Zimov's challenge had to be addressed.

Zimov already demonstrated that grazing animals had a significant influence in keeping the ecosystem stable. Without the herds of mammoths and other megafauna mowing down and trampling plants, shrubs and trees threatened to invade at least parts of the steppe. So when the ice retreated and this productive ecosystem collapsed, did the megafauna go extinct due to loss of the mammoth steppe, or did the megafauna die-off precipitate the steppe's disappearance? The issue is too complicated to deal with in detail here, but a few highlights should be mentioned in Zimov's favor.

From 1927 through the mid-30s, hand formed stone projectile points were found associated with extinct megafauna outside Clovis, New Mexico. The discovery marked the first time archaeologists had proof that people lived in North America late in the Ice Age and hunted large prey.

Perhaps the best description of the weapon associated with Clovis hunters comes from archaeologist Charles C. Mann.

> Clovis points are wholly distinctive. Chipped from jasper,
> chert, obsidian and other fine, brittle stone, they have a
> lancet-shaped tip and (sometimes) wickedly sharp edges.
> Extending from the base toward the tips are shallow, concave

grooves called 'flutes' that may have helped the points be inserted into spear shafts. Typically about four inches long and a third of an inch thick, they were sleek and often beautifully made.[1]

Additional discoveries of these points led to the Clovis First Theory which proposed that Clovis hunters specialized in big game and moved across the land bridge into Alaska down an ice-free corridor and into lower North America where they drove megafauna to extinction.

Migration near the glacial front. (Author's illustration)

Today, as noted in Chapter 37, the theory of Clovis First has been successfully challenged with the discovery of older sites lacking the characteristic Clovis points. It has also been shown that the corridor between ice sheets opened much later than theorized. That suggests that the Clovis culture developed in North America, the oldest site of it being presently found in Texas.

An October 24, 2018 paper by Michael Waters *et al* in Science Advances presented a fascinating discovery in the Buttermilk Creek Complex in Texas.

> "A new point form, a triangular lancet point, appears at the Friedkin site (about) 14 ka (thousand years) ago. This form could have developed in situ from the earlier lanceolate stemmed point and could be the precursor to the lanceolate, fluted Clovis point…Alternatively, the stemmed and lanceolate point traditions of North America may represent two separate human migrations…"

Wherever the Clovis culture originated, the technology used has been closely linked with the timing of megafauna extinction and the collapse of the mammoth steppe. The combined stresses of climate change on the steppe and human predation on large animals with long gestation periods may have doomed the entire ecosystem. However, it is difficult to envision spear hunters getting close enough to inflict lethal wounds on large, fleet-footed animals, and particularly puzzling how someone could garner enough strength to throw a spear into a mammoth's dense fur and fat layer that would render a fatal blow. Dan Fisher (Chapter 8) and other researchers had proven that mastodons and mammoths were part of the human diet, but how were they hunted?

While I was in eastern Montana working on the dinosaur dig in 1992, an archaeologist came into our camp one lunch hour. He carried a spear and a slightly larger diameter length of wood that had been hollowed into a long groove, open at one end and shaped into a raised stop at the other. The device was called an atlatl derived from the Aztec language, and in the 1500s it was greatly feared by Portuguese conquerors, because it could be used to drive modified spears called darts great distances and with enough force to pierce the leather and cotton mail of soldiers.

The archaeologist gave our crew a demonstration. Placing the dart in the channel of the atlatl, he drew back his arm holding the atlatl near its open end. With a mighty thrust the spear shot into the air and landed over fifty yards from us. Cheers went up. We were all astonished. Then we tried with limited success. I could not get the hang of it, but one young baseball player succeeded in throwing the spear nearly double the distance of our archaeologist friend. I had no doubt that it was a superb weapon in experienced hands.

Today, some Eskimo tribes still use the atlatl to hunt seals, while Australian Aborigines use a modified version in selected hunts. It appears to have a long history in deep time. Presently, the oldest example is a reindeer bone version found in northern Europe that dates to 17,500 years ago. In America only younger specimens have been found, but because atlatls are made of organic material, ancient examples are unlikely to have survived the passage of time.

It's tempting to say that the Clovis point was created for big game hunting with the atlatl, allowing high impact, long distance trajectory. But archaeologists need proof. Karl Hutching, archaeologist at Thomas River University in Canada, may have found it.

High velocity impacts on stone create micro-fractures that can radiate from the impact site. Hutching studied such impacts on Clovis points and found that those with damaged tips showed radiating micro-fractures that could be calibrated to the force of impact. His findings in the *Journal of Archaeological Science*, March 2015, show that only the atlatl had the power to deliver such force. Bow and arrow technology that could duplicate such forces developed much later.

It appears that the Clovis point developed as a refinement in technology for hunting big game animals and may have tipped the balance on megafauna survival at the end of the Ice Age. It may have indirectly contributed to the collapse of the mammoth steppe.

A time to trade. (Author's illustration)

42. From Bad to Worse

"Months that should be Summer's prime
Sleet and snow and frost and rime
Air so cold you could see your breath
Eighteen hundred and froze to death"[1]

As the 1816 spring season approached summer in Maine, the weather got no better and became more than an inconvenience. The June 6 snowstorm William M. Brooks recorded in his journal while in Lubec proved to be significant away from the coast. Nine to twelve inches of snow fell that day over the Down East region.[2] At Fryeburg, a western Maine town hugging the New Hampshire border, an unknown diarist described conditions in June. "Frost and ice were common and every green thing was killed. All fruit was destroyed."[3]

As they flew into Maine, migratory birds found a landscape nearly barren of insects and any form of fresh plant life. Depleted of their energy reserves, starvation and cold-shock brought them tumbling from wilted vegetation in the hundreds and thousands.

People picked up the still living jewels – hummingbirds, warblers, martins, and scarlet tanagers.[4] They held them close to provide warmth. Elisha Clark even brought in Baltimore orioles from his China, Maine orchard to warm them by the woodstove.[5] But all the acts of compassion only delayed death a few days at most.

Dying jewels. Author's illustration.

Back in Salem, where the ocean current moderated the weather, the beloved Reverend William Bentley had no understanding of what people far inland and up north were experiencing. On June 12 he penned in his diary "In few seasons have we heard more bitter complaints against cold weather than since June has come in, tho the winter and the whole season, if I may judge from the woodpile, has been as moderate as I have ever known. We shall soon hear complaints of heat."[6]

Five days later a heavy snowfall hit western Maine, and the temperature dropped below freezing.

A farmer, searching for a lost flock of sheep, was out all day in the storm and failed to return at night. He was found three days later, lying in the hollow of a side hill with both feet frozen.[7]

The farmer suffered excruciating pain for many days until gangrene, spreading from his rotting feet, killed him.

When a short stretch of warm weather arrived toward the end of June, farmers first sheared their sheep while letting the soil warm before planting. But cold hit again. Farmers raced to tie fleece back over their newly shorn sheep but couldn't prevent many of them from freezing to death in their stalls. People fell into despair as food supplies dwindled for livestock and themselves. The shrinking hay supply and stalled growth of fresh grass forced them to feed much of their vegetable and grain crops to their cattle. Their Puritan conviction of Providence assured them that warm weather would prevail if they repented and humbled themselves.

Revival spread like a grass fire across New England. Some church congregations grew even without a minister present. But as the summer wore on, people butchered their starving animals as a merciful death and to save themselves. They became so despondent that one elderly farmer who could no longer stand the baying of his starving cattle slaughtered them then hanged himself.[8]

Merciful deaths were not an option for wildlife. Bird bodies littering the ground became obvious to people. The starvation of animals deep in the woods hid the real toll. Each corpse provided a quick meal for other starving animals. Eggs of grouse were found abandoned and woodchucks were rare and lean. Roving bears were common. Even the wary wolves became bold, not only attacking sheep in pastures, but coming into barnyards for them and into the coops for chickens.[9] "It was so bad in 1816 that four Maine townships voted bounties on wolves up to $40.00."[10]

Spectacular sunsets ruled the skies, but their beauty seemed foreboding. Sunspots grew darker and appeared more prevalent. No one fathomed that Tambora's fine-dust cloud veiled the light. The dry fog was attributed to wildfires, grown common because of the lack of greenery. Speculations went rampant about the primary cause. Were the sunspots blocking solar heat, or were Ben Franklin's lightning rods draining heat from the air? Learned men feared that the cold signaled a heat loss in the earth's interior that had been predicted. They could not envision the radioactive realm responsible in maintaining a constant heat at the inner core.

The weather did not grow consistently cold. It seesawed from summer to winter and back. As it warmed, hope rose and people rushed to plant their crops again, only to have a bitter cold wave freeze them as they sprouted. The cycle grew maddening, as though some force teased everyone's sanity. Preachers called it the wrath of God. Others imagined witches had resurrected their traditional arts.

It was not the best time for an elephant to tour the District of Maine.

43. The Mammoth Who Lost Her Steppe

> The termination of the Wisconsinan glaciation has
> been revealed in many parts of the world as a time of
> highly variable climate and vast changes in ice sheet
> masses, oceans and biota. Maine experienced a rare
> combination of important global phenomena, with the
> margin of the Laurentide ice sheet meeting the sea,
> with sea level hanging dramatically as a result of both
> eustatic and isostatic processes and with vegetation in-
> vading newly-exposed landscapes.[1]

This excerpt from a major study that Hal Borns led before, during, and
after the Scarborough mammoth excavation gave an updated, clearer resolution
to the timing and character of glacier retreat over Maine. The study particularly
intrigued me because it put Hairy-it in broader temporal and landscape
context.

When the climate warmed at the end of the Ice Age and the glacier front
retreated over Maine, a major portion of the ice sheet still covered Canada. It
held enough water from ocean evaporation to lower the worldwide coastline by
nearly 400 feet (a eustatic process). But in Maine sea level was particularly
dynamic.

Today the town of Medway at the confluence of the east and west branches
of the Penobscot River lies about 75 miles from the Maine coast, but 13,000
years ago it had an ocean shoreline, because the glacier retreated much faster

than the earth's crust rebounded from below sea level (an isostatic process). But long before that occurred, bedrock emerged very near the present coastline.

Around 17,000 years ago a dark blemish in the ice resolved into a granite island rising above a sea of white. The summit of Cadillac Mountain in Acadia National Park became the first of Maine's land to glisten in sunlight. Interior mountains, much taller than Cadillac's 1,500 feet, still lay entombed in thicker ice much farther from the glacial front. Four hundred years later and slightly over a mile from Cadillac's summit, little Sargent Mountain Pond appeared on the upper shoulder of Sargent Mountain. Today it is regarded as Maine's oldest contemporary body of freshwater.

From the dating of a *Portlandia arctica* shell found at the mammoth site by Hal Borns we now know that the ice left Scarborough 14,800 years ago, while the area was still under water. Less than 300 years later some of the highlands in extreme southwestern Maine protruded from the ice; their dark surfaces absorbed more heat than the receding ice and accelerated the melt-back. Increased summer heat also spawned more abundant meltwater channels that spilled down the glacier front, thinning its face. Ice crevasses became catch basins for surface water to find its way to tunnels at the glacier's base where it propelled a slurry of fine sediment, sand, and gravel, scouring the ice cavities into large arteries bleeding meltwater and stratified glacial till. As the glacier thinned, the bedrock topography of mountains, valleys, and shorelines took on a greater contribution in the fragmentation of the glacial front.

The Bubbles in Acadia National Park appear during late glacial melt back. (Courtesy of the Climate Change Institute, University of Maine, artist Gary Hoyle)

Lakes were impounded in north-sloping ice-dammed valleys, and upstream from temporary drift dams in the Saco Valley.[2]

At that time the Saco Valley served as the major drainage for the Nonesuch River, not far from the Scarborough mammoth site.

By the time Hairy-it grazed across southwestern Maine close to 13,900 years ago, a tongue of the Laurentide Ice Sheet stood roughly sixty miles north of Scarborough at about where Rangeley Lake is today. As land rebound pushed the coastline seaward, water flushing from the eroding ice periodically pooled behind walls of glacial debris until pressure collapsed the rubble and flooded the surroundings. Not all these impoundments collapsed; some remained stable to this day, such as forty-five square mile Sebago Lake, 15 miles northwest of Portland.

As vegetation began invading the newly exposed land, the succession developed similar to that documented in Nova Scotia where a first known vegetative carpet rich in a variety of lichens spread across the landscape. Sedges, shallow rooted grasses, willows and birches eventually crept into the region, persisting as the dominant plant community for two thousand years.[3]

Something similar occurred about this time in the Alaska-Yukon region except that dwarf shrubs of willow and birch still dominate much of the area.[4]

Whatever type of mammoth steppe existed south of the ice sheet, it experienced a similar disruption from shifting climate as that occurring in Alaska and Nova Scotia.

Following game. (Author's illustration)

Human hunting pressure may have compounded the problem reducing the number and types of grazers that encouraged the growth of graminoids (grasses and grass-like herbs) and forbs (flowering plants such as yarrow, mums, tansies and sagebrush). The change would play out directly in the life of Hairy-it, the Scarborough mammoth.

44. Getting to Know Elephant Bet

Elephants are curious and highly intelligent creatures. Early in Hachaliah Bailey's travels with Bet he met a farmer who offered him a two-gallon jug of rum so he and his family could see the elephant. Bailey accepted and stashed the jug in the rear of the trailing wagon he used for supplies at the time. After the show he drew the jug to his lips, while the handler chained Bet to the rear.

That night, as the show moved in a heavy rainstorm, Bailey and his assistants were almost scared out of their wits by a great bellow that set the horses off on a wild run. Old Bet had drained the jug and was fighting drunk. She wrecked one of the wagons and roared and trumpeted all night, and only the heavy chains on her feet prevented her escape.[1]

The following morning Bet sobered and drank gallons of water. After that experience she never showed an interest in hard liquor.

Bet's inquisitiveness made her a quick learner of tricks, but it also led to her developing a particular fondness for Nathan Howes' dog, Buster, as they journeyed along the road together.

Hachaliah Bailey had devised a breakdown shed used to hide the elephant, when no other shelter seemed available. It proved to be an unwilling structure to assemble and had to be transported in a separate wagon. It provided a cozy nook for most animals, but became claustrophobic for an elephant. Over the years Bet learned to endure it. Howes had taken the breakdown shed with him and used it now and then at some of his Roadshow stops. During one of those occasions, when the weather grew hot and people crowded around the shed to see the elephant, some boys began teasing Bet's friend, the dog. With each distressing yelp, Bet responded with a low, guttural moan. Every time the shed door opened to show the elephant, Buster ran inside and Bet lightly stroked her foot over his back, as if to soothe him. When the youngsters realized that they could get such a unique response from the elephant, their teasing increased and some adults joined in.

Confined on a hot day while hearing the sounds of her distressed friend, Bet began to lose patience with this game devised by the crowd. After a series of these incidences, she began swaying aggressively from side to side. Her trunk thumped against the walls prompting gleeful screams from her audience. The hecklers relished the reaction and teased the dog again.

Bet flew into a rage. Again and again she beat the shed walls and with every thud the crowd roared louder. All members of the roadshow hurried to the shed just as Bet coiled her trunk into a ball and hit the walls with all her might, "and the boards flew like feathers…"[2]

The crew terminated the exhibition, calmed Bet, and repaired the shed. By nightfall they were packed up and hurried on their way to the next venue.

Few bridges had been built over rivers and wide streams in the early nineteenth century countryside, making ferry service the common mode of crossing. Horses and wagons routinely rode across the current strapped securely to the deck. For The Nathan Howes Roadshow these crossings were second nature until an elephant joined the troupe.

During one particular ferry crossing with Bet secured to the deck, her friend Buster weaved between passengers. At one point folks heard about the affection shared by the giant and her little pal, prompting some curious people to tease the dog and watch for the elephant's response. Bet groaned and shifted her weight, frustrated that the tormentors lay beyond the reach of her trunk.

But beside her water flowed in abundance. Extending her trunk into the river she sucked up gallons and blew the contents across the boat. Shrieks of laughter erupted from people being doused. Again they prodded the dog and Bet drenched them. The merriment continued until the ferry crew realized the boat was filling with water. As men began bailing, Bet saw it as a new game. For every pail full dumped overboard she replaced it with gallons more. The hilarity now switched to a game of survival. More people began to bail, some using just their hats. Determined to win the game, Bet increased the frequency of her drafts. People did the same. They all worked like a well-oiled machine until the boat thumped into the shore.

Shaken passengers and crew jumped to the ground thankful that the contest was over. As the roadshow crew walked Bet off the ferry, everyone breathed a sigh of relief. Bet walked away perhaps disappointed that she had not sunk the boat.[3]

45. Goodbye Hairy-it

Mammoths behaved similarly to living elephants and, like them, traveled in matriarchal herds. Modern herds are led by the oldest female and consist of females of all ages and only juvenile males. Before they reach sexual maturity males are driven from the herd, to live isolated or in small bachelor groups. Today's elephant herds range from less than ten individuals to as many as one hundred. Mammoth herds likely formed in the same way, though during seasonal migrations herds grew much larger.

During the glacial maximum when the ice sheet reached as far south as Ohio, mammoth herds had a major effect on conifer forests. After fattening on the forbs and grasses of the mammoth steppe, woolly mammoths moved into the bordering spruce-fir forests for the winter, while the Columbians traveled much farther south. Diets shifted to less nutritious browse. Mammoths uprooted saplings and stripped boughs from larger trees for their tender shoots and seed cones. At times they toppled trees to reach the higher succulent browse. Large herds could temporarily devastate a forest, but the abundant dung peppered with undigested seeds guaranteed a constant renewal. The overall effect of mammoths on these forests created a park-like landscape.[1]

As the dominant herbivores of their time, both woolly and Columbian mammoths contributed significantly in maintaining a balanced environment. When the glacier front retreated at the end of the Ice Age, their activities may have been vital in balancing the ecology of the mammoth steppe as it moved northward. But all grazers of the steppes helped in that balance.

Grazers targeted mostly forbs and grasses. Some were generalists, consuming a wide variety of low growing vegetation, while others specialized in restricted diets. Their overall contribution on the steppes favored the spread of deep-rooted grasses and forbs and stimulated rapid regrowth. As a result, water transport due to greater transpiration, aided by katabatic winds, kept soils from becoming waterlogged. This drying further favored the spreading of more forbs and grasses, which in turn supported more grazers to limit the encroachment of rival vegetation.[2] Their manure, laced with undigested seeds spread a riotous summer bloom in new patterns each year.

Mammoths being prodigious consumers of vegetable matter likely played a key role in this feedback loop and in winter kept open the boreal forest that followed the mammoth steppe northward. But rapid climate change severely stressed both mammoths and their habitat. Woolly mammoths were not adapted to the warmer, snowier winters and more humid summers that came. Those stresses likely decreased longevity and reduced birth rate, which posed a major threat to a species with a gestation period of 22 months. Populations of them dropped and the mammoth steppe's composition shifted to a far less nutritious wet tundra plant community. Suitable fodder, rich in nutrients and in high enough density, became restricted to well-drained, windswept, north-slope terrains capable of supporting only small isolated populations of mammoths. These may have become the last mammoth hunting grounds for the Clovis people.

It is tempting to try to resolve these independent studies that I've mentioned into a single scenario that points to mammoths being the key to the mammoth steppe's survival. Sergey Zimov has demonstrated that a high density of large herbivores along with mechanical uprooting of trees, simulating mammoth behavior, can shift at least some areas of tundra into pastureland. Zimov has also compared the productivity of the mammoth steppe with that of the present day African savannah where elephants are the major key for keeping tree growth in check. But the African savannah is a poor comparison to the mammoth steppe, because of the steppe's intimate link to the ice sheet.

If mammoths had gone extinct prior to the ice sheet melt-back, the mammoth steppes would have contracted but not collapsed. Winds would have kept an open prairie near the ice front. Katabatic winds periodically raging across the open landscape would continue lifting billows of the finest rock grindings, dubbed rock flour, from the glacial front and deposit them as loess rains of mineral nutrients over the steppes. Snow, known for trapping a nutrient form of nitrogen, had built up for many thousands of years, compressing as a two-mile thick ice sheet. At its glacial front that nitrogen would continue to be released in fluctuating braded streams and pulses of subsurface seepage at a concentration for optimum plant growth. The combination of these factors would have still created a lush growth of forbs and grasses sustaining massive herds of grazers. Because of these critical features, the mammoth steppe is regarded as a feature unique to the Ice Age.

Looking into deep time is a bit like looking through a plate glass window at a shrub-studded meadow on a foggy day. We can recognize and more easily understand the things closest to us. As we look deeper into the fog, things become less distinct and more open to conjecture. Are those the same types of shrubs we see near to us? Are they growing in dense or open clusters? If a deer herd crosses in the far distance, the faint shadow image may be vaguely understood, partly because of our reflection on the glass.

A great deal of inventive and inspiring work by researchers in many fields over the years has cleared away some of that fog; Dan Fisher's life-studies analyses of ancient elephant tusks is just one example. But a fully clear view of the late glacial world is not yet present. We still look through reflections of suppositions, conjectures and theories that may distort our view of the past. Yet a string of facts can tell a story.

In 2004 a multi-authored paper appeared in the journal, *Quaternary Research*, titled "Late Pleistocene mammoth remains from Coastal Maine, USA." Our team did an amazing job stitching together the intricate science it took to reveal with some confidence what happened toward the end of Hairy-it's life and how her remains became encased in clay.[3]

*   *   *   *   *   *

Nearly 13,900 years ago, a small matriarchal herd of mammoths grazed well north of Scarborough and a few miles south of the glacier's ice front. The verdant slopes above a mud-churning river rippled in the summer breeze. Having scrubbed free from most of their winter coat on spruce bark and boughs nearer the coastline, the elephants relished the breeze blowing the remaining dander from their shorter, thinner fur.

A mature female in the group, our Hairy-it, fed close to the matriarch as though seeking wisdom from the leader, while other members scattered across the hillside grazing furrows through the sward. The old female led her companion to a cloistered patch of buttercups she had found the previous year and let her eat alone. Somehow the matriarch knew this mammoth needed special care. Was it the washboard ribbing on the tusks at the gum line or an odor that gave her a clue that something was wrong?

The herd's recent ancestors had encountered larger, more sparsely haired, mammoths toward the west that bred with some of them. Hairy-it was the offspring of one of those unions. Her molar teeth showed characteristics of both Columbian and woolly mammoths, but the predominant characteristics were from the woolly lineage. Her physical appearance may have been larger than expected due to the effect of Columbian genes, and her coat may have been less dense.

Four years earlier the fields had been richer in forbs. At that time the herd seldom moved far from the area, though it occasionally wandered along the proto-Saco riverbank in midsummer to feed close to the coast where the grain heads of *Spartina* saltmarsh grasses were ripening. Carbon signatures in the last four annual layers of ivory in Hairy-it's tusk have determined that she did periodically consume plants of this metabolic type. Feeding on less nutritious grasses meant building up a thinner layer of fat for the winter. Fortunately, the herd's maritime location tempered the summer and winter temperature extremes and allowed for grazing late into the year.

When the end of autumn neared, cold rains drove the mammoths into the shelter of trees not far from the grassy slopes of the riverbank. The forest became more their home as the season's chill deepened. Snow soon covered what sparse greenery had been available forcing the herd to sweep away white plumes with their tusks to find the precious morsels.

By mid-winter the diet had shifted to mostly browse in the herd's ever-green retreat. Shrubs that had sprouted last spring were ripped from the forest floor. Small trees fell to the mammoths. Limbs bowed by snow were pulled to the ground. The forest was being groomed to maintain its parkland environment.

Thanks to geologist Bob Nelson, Lisa's insect wing mentioned in Chapter 21 was found to be from a species of brown lacewing that inhabits a parkland habitat, giving us a clue to the forest's character at the time of Hairy-it. (Author's illustration)

Sometime in mid to late winter tragedy struck the herd. One member died. It was not unusual for an old leader with worn out teeth to waste away in the impoverished months, nor was it unusual for an inexperienced yearling to be poached by a predator. Yet this was an adult female in what should have been her prime known to us as Hairy-it. Just before the spring thaw, the matriarch led the herd to the riverbank as though to mourn over the mammoth's body one last time.

Death of Hairy-it the mammoth. (Author's illustration)

The carcass had been scavenged over the weeks since Hairy-it's death. One side of the body remained frozen solid. How she died is still a mystery.

46. Old Bet's Final Walk

After sailing east along coastal Maine as far as Belfast in Penobscot Bay, The Nathan Howes Roadshow began its long journey back to Somers. At the mouth of the Kennebec River, the passengers' boat worked against the current and snaked its way forty miles upriver to Augusta where on Wednesday, July 3, 1816, Bet was exhibited at Ancill Kimball's Tavern. That night the Roadshow packed up and traveled about two miles down the River Road to Hallowell for a three-day show at Joseph S. Smith's Washington Hotel. Exhibiting Bet at Kimball's Tavern had been a primer for events during the longer stay in Hallowell celebrating Independence Day. The hope was that people seeing the elephant would spread the news about the remarkable beast faster than any form of advertising.

Since becoming a veteran of the War of 1812, Nathan Howes viewed Independence Day in a more sacred way, and looked forward to the Roadshow's involvement in the celebration hoping that large crowds would attend. The crowds came but so did a sudden shift in the weather.

Temperatures tumbled as a snowstorm hit the area on the fourth. "(I)ce froze to the depth of several inches in shallow ponds."[1] The abnormal cold frightened everyone. Farmers, who had just replanted, looked over their snow-covered fields in shock. Famine weighed heavy on their minds. Then snow fell again on July 6, deepening the gloom.

The Roadshow crew began to worry about traveling the snowy roads, fearing winter returned to stay. But Nathan's greatest concern focused on the health of Bet. If the Crowinshield elephant did not survive the winter in Boston, could Bet survive the sudden cold? She had acclimated well to overwintering at Wetmore's Inn in Albany, New York, since 1814[2] but that was partly due to heat infiltrating the stable from adjacent visitors' quarters. Could she handle the shock of a rapid chill? Only time would tell. Since his New Bedford encounter with Hachaliah Bailey, Nathan had been punctual with his payments and wrote to Bailey of his progress when he could. Whatever concerns he shared with Bailey are now lost, but sometime during the Roadshow's journey Hachaliah shared a plan with Nathan about what to do if Bet did not survive her trek through Maine.

In a light snowfall The Nathan Howes Roadshow traveled out of the river valley late on July 7th heading ten miles northwest to Readfield Corner. There Bet and the crew entertained a crowd at James Fillebrown's Inn the next day despite it still being unusually cold.

From Readfield they went six miles south to Winthrop where Bet was exhibited at Dean Howard's Tavern on July 9th when snow fell again and Professor Parker Cleaveland recorded a temperature of 33.5 degrees F. on his thermometer at Bowdon College in Brunswick, about forty miles closer to the coast.[3] That night the Roadshow journeyed another six miles south to Sewall Prescott's Tavern in Monmouth for a show on the 10th.[4]

Leaves hung brown and sparse on trees, and cold settled in a dry volcanic fog over the landscape. When snow melted back, patches of green became mostly restricted to fields of rye, wheat, and other cold hearty grains. Corn, fruits and vegetables were near total losses. Along with the cold, few rains came that year prompting a report that, "Unless we have rain soon and constant warm weather, the crops must be very short."[5]

Wildfires propagated as a dull palpable fear seeped into the land. People craved relief from their despair. Some families left the District of Maine for what then was known as the west, most notably the State of Ohio, but the flood of migrants would go the following spring, when weather conditions in

Maine still remained foreboding. Up to 15,000 people left the region smitten with "Ohio Fever."[6] A few things could momentarily break the despair of those who stayed behind: hymn sings, family gatherings, barn dances, and a wondrous elephant touring the District.

Fourteen miles south of Monmouth, on the east bank of the Androscoggin River, The Nathan Howes Roadshow concluded a successful performance in Lewiston, a town then known as Lewiston Falls. At about the same time the *Boston Gazette* quoted a farmer in Kennebec Valley, roughly twenty miles east of Lewiston, whose crops had been killed by the latest frost.

"A famine for man and beast, seems to stare us in the face."[7]

Nathan planned to continue on a southwesterly course and enter New Hampshire at South Berwick, Maine, but first the Androscoggin River needed to be crossed.

At nearly 180 miles in length and draining over twenty significant tributaries, the Androscoggin River is the third major waterway in Maine. Lewiston is less than 30 miles upriver from where it empties into Merrymeeting Bay, and in 1816 a single log dam controlled the water flow only to a minimal extent. Raging currents ruled, but in July of that year drought had slackened the flow and two competing ferryboats glided unhindered across the river.

Paul Hildreth's oversized flat-bottom boat still traversed the water about half a mile below Great Falls, where the river flowed fast but predictable. Hildreth prided himself for being Lewiston Falls' first white settler, moving his family there in 1770. A fire that autumn burned their newly built log home to the ground forcing the family from the area, but the Hildreths returned the following spring and rebuilt the cabin while providing a ferry service for new arrivals.[8] Over the years ferry duty passed to other skilled navigators in the family. But by the early 19th century the need for commercial transport had outgrown what ferries could provide. The population had also grown to a little over a thousand. It was time for a bridge.

Construction to span the Androscoggin began in the spring of 1807 at a little over a quarter mile above Hildreth's crossing. Late summer the following year as the bridge neared completion, a squabble broke out over which side of the river would manage the toll. Not until 1823 was the question resolved and a booth placed on the Minot[9] side officially opened the bridge to public access.

In 1812 while arguments raged over the tollbooth, Zebina Hunt in Minot saw the opportunity to run another ferry service.[10] Using the bridge pilings as attachments for his cable, he could work his way with confidence across the current.[11] Though it broke the monopoly the Hildreths had on ferry service, Hunt's enterprise also took pressure off it. Both ferrymen knew their extinction was imminent.

By the time The Nathan Howes Roadshow prepared to cross the river in 1816, the troupe faced an unfinished bridge. If the structure was passable at that time, the crew may have been able to travel across it. More likely, the Hunt and Hildreth boats ferried freight, horses, small animals and people across—but not an elephant. Nathan Howes was unwilling to give Bet another chance to play "sink the ferry boat." Letting Bet swim the river seemed the only practical and safe recourse.

When the elephant handler reached the opposite shore, some of the remaining crew led Bet to the water. At the handler's call, Bet marched into the river and began swimming toward him. But in midstream she stopped as a spectator remembered, "…and so pleased was the elephant that she refused to come out of the water for a long time. She would swim around and squirt the water in all directions." [12]

A faint blush had grown in the evening sky when Bet came to the keeper's side. The jubilant crowd that had gathered on both shores quieted and dispersed. People headed home for a late meal. Draped in her blanket, Bet trailed the Roadshow with Buster and the keeper as they headed southwest in the growing gloom.

*   *   *   *   *   *

By July 24th The Nathan Howes Roadshow performed in the little town of Alfred, over 60 miles from Lewiston. The grueling tromp through the District of Maine was near the end, 20 miles from the border of southern New Hampshire.

47. A Place Down River

As the spring weather advanced in southwestern Maine a little less than 13,900 years ago, the mammoth herd wandered across well-drained, south facing slopes where the earliest grasses sprout. Ice had broken up on the river, but downstream it formed a jam that flooded to the upper banks, submerging the remains of Hairy-it. When the jam broke, current sucked the frozen carcass into the mainstream and flushed it down river. Braided streams feeding into the channel diverted the body into eddies where ravens pecked the thawing flesh.

Rains came. More water sheeted off the glacier and festooned from tunnels at its base. A muddy torrent wrenched the carcass into midstream and swept it away.

The river widened and deepened. Tusks kept the buoyant body from surfacing. By now the flesh had thawed and fish ripped into it as the river's turbidity lessened.

When the current subsided salinity rose rapidly, as the carcass snaked through the coastal estuary near the river's mouth. Offshore waves raced up the channel, disrupting the flow into whirlpools radiating turbulence that spread the scent of flesh. More fish congregated. The carcass slipped into a bay where silvery shoals condensed from the murky water and flashed across the dark woolly blot, making parts of it tremble with delight. Then in a flash the scintillating mass left, drawing a cue from a dark form rushing toward the corpse. Rows of serrated teeth dug deep as the great white shook chunks free. (A scar on the atlas bone has been interpreted as a tooth mark from a scavenging great white shark.) Again and again the shark tore away flesh. The left side ripped away with a tusk attached, and the incoming tide pulled it free of the violence. It settled in quiet water where the tusk anchored in the muddy bottom twenty feet below the waves.

Lengthening days brought the sun higher and gave it more time and power to bake the dirty glacial front. Gray meltwater bled in ruptured spasms as the ice wall collapsed into withering fingers retreating north. Mud and gravel spilled over the proto-Nonesuch estuary. Pressure waves whipped the nearby seafloor into a muddy cloud. The mammoth carcass rocked but stayed moored by the tusk; rock flour lifted around it leaving only the heavier sand to resettle. A cyprid, the settling stage of a barnacle, cemented itself to one of the ribs and began its transformation to adulthood.

Shifts in tidal currents brought quieter waters over the mammoth remains, where muddy loads settled. Autumn storms sifted out sand. Though many of the skeletal elements had disarticulated, they remained in close association locked in fine sand and the mud beneath them. By winter the bay had frozen over, allowing the finer mud to settle over the bones and smother the barnacle. The return of spring opened the bay again to wave action and meltwater intrusion.

Each year a layer of sand then a layer of mud accumulated. The mud eventually turned to clay. The bay grew shallower and the shoreline advanced due to the continual crustal rebound. Wave base disturbance left its mark as sand ripples in the clay. Mud washed from shallows into deeper water. Tidal surges, waves, and the influx of freshwater reworked what was becoming a sand beach. An avalanche far inland changed the river's course, so the mouth emptied seven miles farther to the southwest. The diminished waterway would eventually be known as the Nonesuch River, and in 1959 two men digging a farm pond in a shallow tributary of that river would find the tomb of Hairy-it.

48. Adding to Despair

Much of the following dramatization is rooted in the recent research by Bruce Tucker, President of the Alfred Historical Society in Alfred, Maine.[1]

On July 24, 1816, twenty miles from the New Hampshire border, Daniel Davis Jr. stood in line outside a tavern stable in Alfred to see the elephant inside. It's doubtful he waited alone. His brother David had died ten months earlier, leaving Daniel with the responsibility of supporting two families living on his brother's farm. Years before the war he and David had built a sawmill on the Middle Branch of the Mousam River in Alfred. The business became so successful that both men borrowed substantial money for a major expansion. But the war hit, and an embargo diminished production to the point that by 1816 David's land was scheduled for auction later that year, despite Daniel selling his own farm to fend off creditors. The oscillating weather extremes amplified Daniel's history of loss and despair and gave little hope for the family's future together.[1]

Word traveled down from Waterboro, four miles to the north that a wondrous beast was coming, something never seen in the District. To watch such an animal perform would be one last hurrah for the Davis family before the auction, but whether the family or just Daniel saw the elephant is not known.

Of the many tricks Bet had learned perhaps the most popular centered on ginger biscuits, which could be purchased in the display barns. It became customary for people to place these treats in their pockets and approach the elephant. Bet then fished for gingerbread with her trunk, pulled the treats from pockets and swallowed them to the great delight of the crowd.[2]

But Daniel decided to use a wad of tobacco as a substitute. Bet had tasted tobacco in the past and had reacted violently. One recent incident occurred in the stable-yard of Timothy Boston's Columbian Tavern in Portland, Maine. As a twelve-year old boy stood awestruck before Bet, he later recalled, "I saw a man wrap some tobacco in paper and give it to the elephant."[3]

Bet had become so notorious at snatching papers from pockets and making a snack of them, that people were warned to remove valuables before testing the elephant's ability to find gingerbread. So it was no surprise that Bet took the package without hesitance from the man in Portland and shoved it into her mouth. She bit down, shook her head violently, and spit the wad to the ground, causing the man and some of the crowd to break into laughter. Bet coughed and flung saliva. As shrieks erupted in the crowd, the trainer ran into its midst and discovered what the culprit had done. When the man insisted that it was a harmless prank, the keeper warned him to leave immediately, because when Bet recovered from her bout of coughing she would grab and kill him if she could. The man sobered but did not leave until the keeper's warning turned into a plea for his safety.[4] Earlier incidences of this behavior may have been the reason Bet's tusks had to be cut back to more safely control her if a repeat encounter of this type occurred.

Once Daniel got into the barn a girl approached him selling ginger snacks for the elephant, but he turned her away. He shuffled through the line along a boarded wall and heard the gleeful shouts on the other side. A few people spoke to him but he felt too depressed to say much. He kept his head down and slouched along toward the glow of lantern light. Swallows flew above him chattering as they darted in and out of the gable barn window. But all the sounds seemed to disappear as Daniel fingered the boards on the wall. In the corduroy pattern of saw marks a repetitive deviation appeared that he recognized. The chipped tooth on a saw blade had caused it: the blade his brother,

David, had changed at the mill just a few years earlier. He could not take his eyes off it, until someone nudged him from behind and he turned the corner to face a most amazing creature. Behemoth, beast par excellence of Holy Scripture, stood before him.

Bet rifled her trunk into the pocket of an adolescent boy just ahead of Daniel. As she plucked gingerbread from its recess, the boy erupted into a nervous giggle then walked away. Now Behemoth faced Daniel.

With a slight tremble Daniel stepped forward. The keeper parroted his warning about tobacco near the elephant, but in his daze Daniel couldn't hear it. He pulled a wad from the pouch in his pocket and squeezed it in his fist. Lifting it to Behemoth, his fingers, like petals, spread to reveal the offering at their base. Behemoth reached out to accept it.

WHAM! The heel of a hand hit hard across Daniel's wrist. Tobacco flew in all directions. Daniel recoiled in pain.

"I said no tobacco!" yelled the keeper.

Daniel's first instinct was to hit back, but his wrist hurt too much. He cussed at the keeper claiming he meant only a little harmless fun, but other members of the crew arrived and drove him from the barn. Humiliated and enraged, Daniel stomped away, determined to get revenge.[5]

*   *   *   *   *   *

By early evening the show was packed up and heading for Sanford and North Berwick: the last stops before New Hampshire. The weather had warmed, and four days of rain in mid-July greened the replanted landscape giving hope for a late harvest—even the moon and stars shown brighter. After the challenging weather, Nathan looked forward to leaving the District of Maine far behind him.

A cluster of about twenty town folks, young and old, journeyed with the caravan, chattering among themselves and the crew, with the intent of wishing them well on the Alfred-Sanford border at Hay Brook bridge a little over two miles down the road from the village square. Thrilled to be walking with an elephant, the locals yielded ample space for Bet to trail untethered behind the wagons with her trainer and Buster the dog.

As they approached the halfway point bordering the plains where a harvest of rye looked promising, Nathan glanced into the trees at the shimmer of an expansive spring-fed pool known as Round Pond. An animal seemed to move in the bushes.

BLAM! BLAM!

A pair of musket balls zinged by the performers. Women and children screamed. Some men crouched low, scanning every recess fearing Britain had revoked its treaty and launched an attack. Bet lunged forward and started down the road faster than she had ever run. She raced by the startled horses spewing blood onto the road from wounds just behind her shoulder blade. Buster and the black handler ran after her. Bet began to slow and weave. About fifty yards down the road she toppled.

When the keeper got to her, Bet lifted her head but did not struggle. Buster danced nervously around her yipping and whining. Dark hands gently stroked Bet's neck and face as she dropped her head. Her heavy breathing eased. The keeper's soft rhythmic voice whispered into her ear. A ruddy-brown eye followed the African's moves. A chorus of mourners assembled behind him, but the eye stayed fixed on the black man. Breathing shallowed to a rasping murmur then died as the eye dilated to a black beyond midnight.[6] The keeper's bitter cry echoed off the hills.

The death of elephant Bet. (Author's Illustration)

*   *   *   *   *   *

Secrets are hard to keep in a town of roughly 1,200 people, but Daniel Davis Jr. thought his was safe. He felt good about killing the elephant. He had paid money to see Bet and got cheated. Now the elephant owner couldn't cheat anyone else.

The town was all abuzz, but the conversations were not in Daniel's favor. When the news hit the papers, Daniel's elevated mood crashed and he went into hiding. The local press fully described the incident, calling the perpetrator an "unprincipled villain." And adding: Endeavors have been made to discover the author of this base act.[7]

The news did not stop at the District level nor did the outrage. In Salem, Massachusetts a newspaper stated: The vile act in destroying the Elephant which has been exhibited in the United States to gratify the innocent curiosity of our citizens, has a full share of the public abhorrence.[8]

Nor did it stop in New England as noted later that year in the *Boston Gazette.*

> Our Alfred Correspondent's letter, giving an account of the death of the Elephant, has been published in almost every paper in the Union...[9]

Reader interest led to republishing in pamphlet form the letter along with a lengthy opinion and descriptive piece by *National Register* editor Dr. Mead.

By the time Daniel Davis Jr. was arrested on September 5, the press had labeled him: "...the vile monster..."[10]

His defense amounted to the claim that exhibitors had taken money from people who could least afford it. It echoed his deep despair. Bail was set at $500.

49. A Double Resurrection

Shortly after Bet's death Hachaliah Bailey's plan for the elephant's remains went into effect. Bet had grown to be a legend and Bailey wanted that legend kept alive. He knew that some of Nathan's relatives in the Roadshow prided themselves as skilled butchers, selling their meats at the family store in Brewster. Now, he needed those skills applied. He would pay the crew and experienced butchers needed from the town to skin the elephant and deflesh the bones immediately. A team of oxen brought to the site sledged the body to a suitable location for butchery. Soon the first shift of skinners began their work under oil lamplight. By morning the hide had been pulled off half the corpse and a new team replaced the bloody, exhausted crew. Early on, the viscera had been gutted from the body cavity and hauled away, lightening the carcass and allowing it to cool more rapidly. Butchers detached the skinned legs and quickly sliced meat from the bones while horses hitched to elephant legs on the recumbent side rolled the body so skinning could continue. By noon the job had finished. Bones and the salted hide were crated and on a wagon hurrying out of the District while the destination of the sacred flesh remained a well-kept secret in a region on the fringe of starvation.[1]

When Nathan reached New York City with Bet's remains, he headed to one of the major tanneries in the Bowery where he met Hachaliah Bailey. Bailey had been shaken by the death of his beloved elephant but now his focus centered on resurrecting Bet. He knew many of the tannery men from his dealings at the Bull's Head Tavern and had already negotiated the best treatment for Bet's hide. He also knew the best man to clean bones, articulate the skeleton, and accomplish the rare feat of creating a taxidermy mount of Bet.

The most notable taxidermist in New York at that time proved to be John Scudder who then owned the American Museum. He also had a great deal of experience cleaning and articulating skeletons, having worked on museum specimens since 1802 when Edward Savage hired him as the keeper of his Columbian Gallery of Painting and City Museum. Without a doubt Scudder was the most skilled and trusted person to work on the remains of Bet.

While fleshing knives shaved extraneous tissue from every square inch of Bet's hide in preparation for tanning, John Scudder's covered, warm-water vats leaked the stench of putrefaction from bacteria feasting on the surface residue of elephant bones. Months later after the whitened bones dried from a bleaching bath, Scudder and his assistants would join them together on a wrought iron framework to create an upright skeleton. From that he would estimate the size and proportions of a mannequin needed to create a glove-like fit for the skin and give the feeling of life to the final mount. But in the meantime he had a problem to solve that vexed him for over a decade.

Scudder felt that the eye functioned as a window to the soul. Over the years he had seen too many animal mounts suffer dead expressions simply because of inferior artificial lenses. The problem often kept him in his workshop late into the night experimenting with a range of ideas. Now tasked with resurrecting a celebrated elephant, John Scudder redoubled his efforts to make the perfect eyes while bacteria gobbled tissue off Bet's bones.

*   *   *   *   *   *

Being a successful cattleman Hachaliah Bailey knew the importance of good husbandry. Crops could fail; disease could devastate a herd. Frugal living helped, but investments and insurances provided the firewalls against failure, and with the success of exhibiting Bet, Hachaliah had the cash for both.

With Bet's death Bailey had lost three thousand dollars of show money a year, but the elephant was insured. Not long after he had organized plans for Bet's resurrection with John Scudder and the tannery, Hachaliah received his insurance money - $30,000 (nearly half a million dollars in today's money). Papers trumpeted the news, sending a shock wave through Putnam and Westchester counties. People of the region knew that show animals were moneymakers, but not to that extent. Bailey's neighbors wanted part of the action. But Hachaliah stood well above the crowd, with capital and an experienced trainer. He had also partnered with his brother-in-law, Isaac Purdy, and a distant cousin and neighbor, George Brunn, to work out details of a tour before allowing Nathan Howes to lease Bet. Brunn came from the family that had worked with the elephant at least since 1808. Now time was the issue to fill the vacuum left by Bet's death.

Having no time to devote to the trial of Daniel Davis, Hachaliah never pressed charges on the destitute man. Soon Bailey and his partners invested in two elephants to replace Bet: this time dealing with ships headed to Bengal.

* * * * * *

On April 5, 1817, the following ad appeared in the *New York Post*.

The Skeleton of that unfortunate Elephant shot last July 26th
in the District of Maine, so well known to the public, is set
up for inspection and may be seen at No. 301 Broadway, from
Monday the 7th until Wednesday the 30th instant.

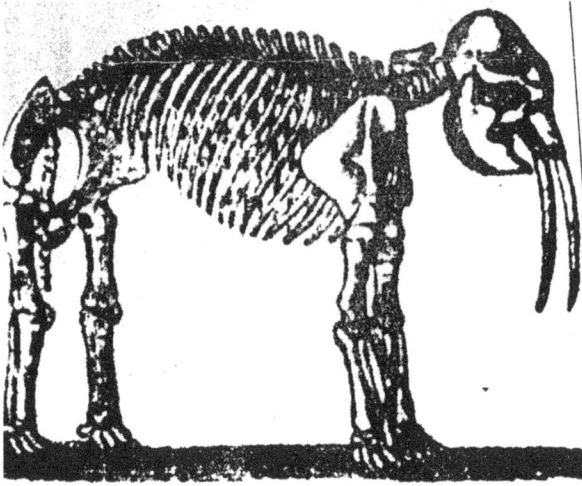

Notice that according to this illustration, Bet's tusks were mounted in reverse of what they would have appeared in life. This is not because of the ignorance of the taxidermist. Mounting them in a lifelike position would have overstressed the skull, fracturing or completely destroying its frontal aspect because of forces resulting from the cantilevering of the tusks.

John Scudder had worked frantically through the fall and winter to meet Hachaliah's deadline for the showing of Bet's skeleton and lifelike mount in the spring. But when the weather warmed, only the skeleton stood ready to show. When Bailey saw it he wanted the impressive piece on exhibit without waiting for the taxidermy mount to be paired with it. Later that year when Scudder finished his elephant, Bailey chose to exhibit it separately in another part of the city. Bet's double resurrection proved to be enough of a moneymaker for Bailey to turn a profit until the live elephants arrived. It wasn't until late November that a newspaper notice appeared regarding his purchase.

"The public are informed that this morning was landed the long expected FEMALE ELEPHANT, and is to be seen at No. 296 Broadway..."[2]

Hachaliah christened her Little Bet in honor of his daughter and his fateful elephant now called Old Bet. Sixteen days later a male elephant arrived in Boston.

"The Elephant COLUMBUS, to be seen in the avenue opposite the Old South Meeting-house."[3]

Like her predecessor Little Bet took to training so easily that she became known as The Learned Elephant. She toured widely at times with The Nathan Howes Roadshow and became a popular attraction. But on the night of May 25, 1826, Little Bet met the same fate as Old Bet when a group of ruffians gunned her down as she crossed the Chepachet Bridge in Rhode Island.

Columbus survived until 1851 when he fell through the floorboards of a bridge in North Adams, Massachusetts.

50. End of Another Legend

As the third decade of the 19th century approached, both the Croton and Danbury-Peekskill Turnpikes swelled with travelers. The intersection in Somers became a popular rest stop, but the old Leggitt Tavern needed a facelift to compete with other local establishments. Hachaliah Bailey decided to raze the structure and build a grand hotel. He broke ground for it in 1820. By then his investments from exotic animals, riverboat services, cattle and many other ventures had made him a rich man and a local legend. But he credited the budding of his success to a single elephant. Though Old Bet's remains no longer made much money, they had become in a sense sacred.

In August of the following year this notice appeared in the Ladies Literary Cabinet.

DIED,

"On the 7th inst. after a lingering illness, John Scudder, Esq. proprietor of the American Museum, in the 45th year of his age."[1]

Like all taxidermists of his day, Scudder worked with compounds of mercury, arsenic, and a host of other lethal substances in order to preserve his specimens. In addition his maceration vats, for removing extraneous tissues from bone through putrefaction, harbored potentially dangerous bacteria. Life was short for many of these practitioners.

That December the American Museum placed an ad in the New York Post displaying the image of an elephant and stating:

Within the last six months, vast additions have been made to this establishment in its various departments, of which it may be proper to mention a few, - of these the most remarkable is the Elephant, which was imported into America in 1804, and after being viewed with wonder and astonishment, by our fellow citizens, for about 12 years, was killed by a barbarian, in the town of Alfred, County of York, and State of Maine, 24th July 1816.[2]

The ad went on to state that the elephant "…is placed, in such a manner, as if it were really alive."

The American Museum seemed the obvious proper repository for Hachaliah's beloved elephant, but with John Scudder's unquestionable talent the timing of the gift speaks also as a memorial to the taxidermist's art.

*   *   *   *   *   *

Bailey chose for his hotel a design that today is considered a "rare, distinctive example of Federal Period domestic architecture."[3] When he opened it for business in 1825, its name was emblazoned across the bricks above the second floor – ELEPHANT HOTEL. The three-story structure made a stately addition to the town and became known as the finest hotel between New York City and Albany.

At the front of the Elephant Hotel, Bailey erected a granite pillar in 1827 crowned with a small, hand carved, wooden elephant covered in gold leaf. Some people speculate that it is a memorial to Old Bet, but Little Bet had died the previous year, so it may have been meant for both elephants. Others suspect that the bones of Old Bet lie buried beneath the monument, since no record survives of the skeleton's fate. If Hachaliah buried the bones, he likely did so discreetly to avoid pilfering by souvenir hunters. Old Bet, by then, had become a legend.

The following year, while in Bethel, Connecticut, Bailey happened upon a recently opened confectionary and fruit store run by an affable eighteen-year-old proprietor. The establishment was rapidly becoming a local gathering place where beer and oysters nurtured hearty conversation, pranks, and hilarity. Hachaliah felt at home and, when in the area, visited the store as often as he could. He loved to reminisce about his escapades with Old Bet, and each tale enthralled the young storeowner, Phineas Taylor Barnum.

As the 1820s came to a close, so many people in Somers and the surrounding towns entered the menagerie business that the competition became unruly. An agency seemed necessary to control travel patterns across the eastern U. S. to reduce conflicts arising over venues. The solution to this became formally addressed on January 4, 1835 in a meeting at the Elephant Hotel. That day a hundred and thirty people involved in the menagerie trade formed the Association of the Zoological Institute.

> The Zoological Institute became a dominating factor in the importation and exhibition of animals and the ownership of circuses and circus equipment. It maintained a collection of animals, wagons and equipment and owned a building at 37 Bowery where performances were held and winter quarters maintained. It organized the numerous menageries into twelve companies for which it established fixed, non-overlapping routes throughout the eastern United States. The Association exercised a virtual monopoly on the animal show business.[4]

However, two years later the Association dissolved, when a deep recession hit the country. On August 22 and 23, 1837, to pay off debts, an auction of animals and equipment took place at the Elephant Hotel.

In his autobiography of 1855, P. T. Barnum described "Hack Bailey" as a self-willed man. When his mind was made up, there was no turning him. The fall of the Institute and the resulting power plays affected Hachaliah Bailey in such a way that within a year he sold the Elephant Hotel and moved to Virginia where he lived out the rest of his days.

Portrait of Hachaliah Bailey
(Courtesy of the Somers New York Historical Society)

In 1845 he returned to Somers to visit relatives and friends. While there he got kicked in the head by a horse and died a few days later. He was buried in Somers with the stone inscription "Enterprise, Perseverance, Integrity."[5]

51. Drawing the Final Curtain

When the taxidermy mount of Old Bet arrived at the American Museum in 1821, five museum trustees had taken management of the establishment. John Scudder specified to it in his will. His oldest child, John Jr. was only 14. However, the boy displayed a precocious curiosity and intelligence. Two years later when the museum published a guidebook, the trustees included John Jr.'s descriptive notes about the collection. The museum kept to its traditional goal of being an institution of public learning and received praises by Gideon Welles[1], who would become one of the founders in 1842 of the oldest continually operating public art museum in the country, the Wadsworth Atheneum in Hartford, Connecticut. But this noble purpose became assaulted from inside the institution and out.

John Scudder, Jr. began quarreling about the museum with the trustees and his four sisters. Finally, in 1825 after dropping out of medical school, he opened a rival to the museum that he titled Scudder's New York Spectaculum. That autumn another more potent rival appeared, the Peale's New York Museum.

Rubens Peale was one of the sons of Charles Willson Peale, creator of the celebrated Peale Museum in Philadelphia. Unlike his father, whose mission was to educate the citizenry, Rubens intended to please the visitors stating:

"…if they are gratified, they gratify us with their money."[2]

To that end, he primed himself as a showman, while retaining the skills of museum promotion that his father taught him.

216

New York City grew rapidly but had not yet reached a population that could support three museums with similar competing goals. John Scudder, Jr.'s Spectaculum became outmatched and soon closed its doors, leaving Peale to compete with the American Museum. In their competition for dollars, both museums shifted their focus toward entertainment, resulting in a gradual lowering of the quality of amusements. In 1830 the rowdy crowds drawn by the performances at the American Museum contributed to the eviction of the museum and other civic establishments from the New York Institute. The museum lost its rent-free status and moved into a leased building, hiring John Scudder, Jr. to manage the institution. Scudder, by then, had slipped into debt and alcoholism, but he retained the showman's charisma he had honed at the Spectaculum and soon put the museum into the financial "black." Scudder's appointment didn't last long however. In January of 1831 the trustees fired him after his involvement in a drunken brawl.

The next several years challenged both New York museums. First, a cholera outbreak in 1832 diminished visitations, then a fire swept the city in 1835. But the coupe de grace for Peale's New York Museum came from the deep financial recession of 1837. The American Museum barely survived. Once again John Scudder, Jr., returned and pulled the museum out of debt with his showmanship. This time the lack of major competition assured that the museum would survive. Healthy profits in both 1839 and 1840 stood at over $11,000. But everyone associated with the American Museum had grown weary, and in 1841 the institution went up for sale.

By then the young storekeeper, enthralled with Hachaliah Bailey and his tales of Old Bet, had reached his early thirties and had tales of his own. The Barnum store had been a local gathering for good-hearted pranks and tall stories. Phineas relished that atmosphere, having grown up with it while clerking as a young teen at his father's store, and he recalled

"... our store was the resort of all these wits, and many
is the day and evening that I have hung with delight upon
their stories, and many the night I have kept the store open
until eleven o'clock, in order to listen to the last anecdote..."[3]

From tending his own store and publishing a small town newspaper, Barnum learned that good humor and aggressive advertising proved to be a successful mix. He tried his hand as a showman, beginning in the New York area, where he learned the art of mild, "tongue in cheek" deception, and loved it. But his travels with a circus did not suit him. He was a gifted businessman yet had gone through pendulum swings from comfort to despair on many occasions. By the time the American Museum went up for sale, Phineas Taylor Barnum had become essentially penniless with a family to support. Yet he had big dreams, and he could see potential in the American Museum stating:

...I repeatedly visited that Museum as a thoughtful looker-on.
I saw, or believed I saw, that only energy, tactic, and liberality
were needed, to give it life and to put it on a profitable footing...[4]

Through stellar references, strict financial agreements with the building's and museum's owners, and a number of fortunate twists and turns, P. T. Barnum became owner of the American Museum on December 27, 1841.

John Scudder, Jr., had already proven how vital showmanship could be for the museum's survival. Barnum directed such entertainment to the Victorian sensibilities of the growing middle class and avoided the more tawdry acts that Scudder had promoted. Over a decade later Barnum remarked, "The *moral drama* is now, and has been for many years, the principal feature of the Lecture Room of the American Museum."[5]

However, his targeting of the middle class had little to do with moral concern. He reasoned, "The only way to make a million from my patrons was to give them abundant and wholesome attractions for a small sum of money."[6]

To induce them to frequent his establishment, he also introduced freaks of nature into his exhibits, some manufactured from body parts of different animals. He claimed his humbuggery was meant to question, entertain, and ferret out the truth, much like the practical jokes of his youth. In his defense he wrote, "I cannot doubt the sort of 'clap-trap' here referred to, is allowable, and that the public like a little of it mixed up with the great realities which I provide."[7]

Those great realities numbered about 50,000 when John Scudder, Jr., wrote his guidebook for the museum; under Barnum they rose to over three-quarters of a million. Of them Barnum had an old favorite. At a bail hearing in support of a friend he listed that item and its value as number one.

"One preserved elephant, 1,000 dollars." [8]

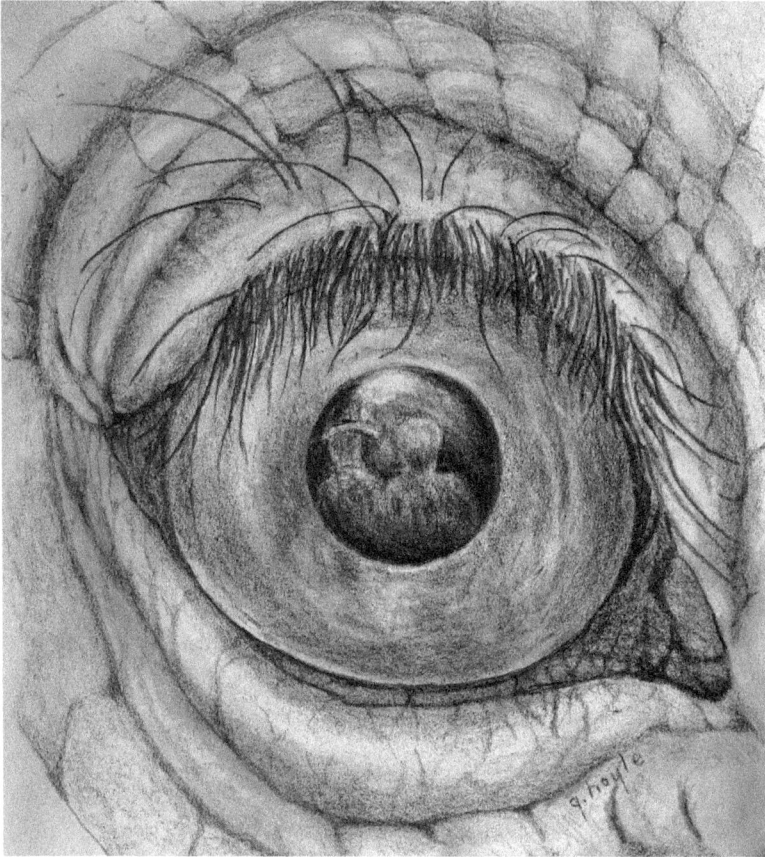

Merry reflections in Bet's glass eye (Author's Illustration)

Old Bet was Barnum's touchstone to the cherished times of his early life—the jokes, pranks, wild tales and stories of adventure that inspired him to be a showman. He once said, "I feel myself entitled to record the sayings and doings of the wags and eccentricities of Bethel, because they partly explain the causes which have made me what I am." [9]

A year after taking charge of the American Museum, Barnum became debt free. Within twenty-six years under his leadership the museum stood as the major draw for entertainment in the city. At its peak it employed a hundred and fifty people and was open fifteen hours a day. It became the Mecca for tourists and dignitaries as well as locals. Between 1842 and 1865 an estimated 38 million visitors paid the twenty-five-cent charge to spend a day in the museum. Put in perspective, the population of the country in 1860 stood at less than 32 million.

But in one day, July 13, 1865, it all literally came crashing down. Fire erupted in the lower part of the building. Fortunately, no people were killed in the blaze, but a host of caged animals burned alive or were shot escaping the inferno. It remains one of the worst fires in New York's history.

Little of value survived. The part of Old Bet that remained in the limelight twenty years after Hachaliah Bailey's passing had become ashes. But Barnum prevailed. He opened another museum that same year, but it too burned in 1868. He then went into politics and the traveling circus industry. In an odd twist of fate he eventually teamed up with James Bailey, the adopted son of a distant cousin to Hachaliah Bailey. The Barnum and Bailey Greatest Show on Earth would be one of the first and most successful circuses to travel by rail. It eventually travelled worldwide.

Bet's charred and shattered remains (Author's Illustration)

52. Old Bet Remembered

On July 24, 1963, a crowd of people gathered near Round Pond in Alfred, Maine, at the spot where Old Bet died. Associate Supreme Court Justice Cecil Siddall of Sanford gave a brief history of the elephant as was understood at that time. The Sanford-Alfred Historical Society had arranged for a dedication to Old Bet on the 146th year of her death along with the Circus Fans of America. At that time Mrs. Logan Billingsley, vice president of the Somers, New York Historical Society and director of the Circus Fans of America, unveiled a bronze plaque affixed to a large granite boulder commemorating the site of Old Bet's death.[1]

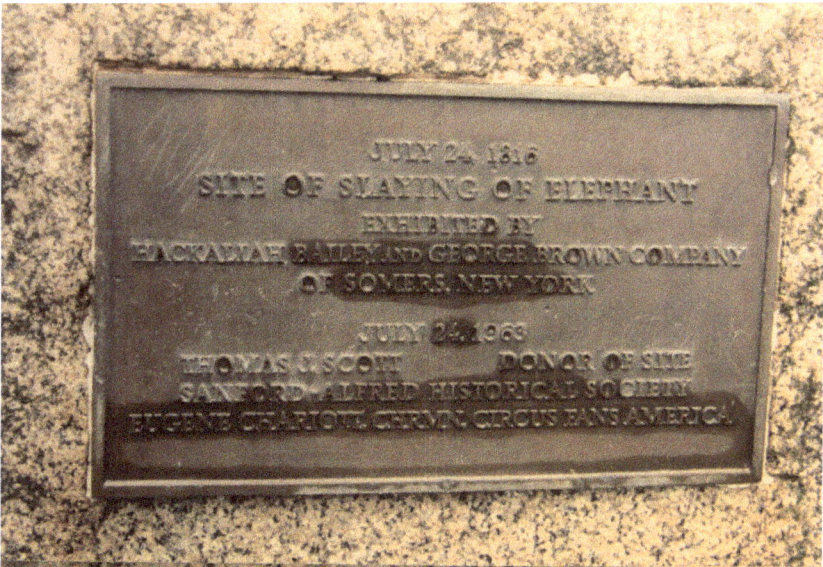

Old Bet's memorial plaque in Alfred, Maine (photo by the author)

Over thirty years later I stood at the monument and tried to envision the area as it appeared in 1816. The weather had turned from showery to slow clearing, but tires hissed as traffic raced by. The road had become a major thoroughfare and houses dotted the landscape; fields had grown into woods and suburbs. Below me, off to my left, lay the little body of Round Pond that, through the cluster of budding spring branches, appeared to be surrounded by a chain linked fence. Without the bronze plaque, nothing in the landscape could provoke a touchstone to that time so alien to our fast paced world today. I left with not a vision, but many questions. Perhaps that's the point of the monument – to wonder and want to know more.

53. Personal Thoughts

Two elephants died in Maine over 13,600 years apart. They became connected only by legend, supposition, and deception. In 1959 Christopher Packard, Director of the Portland Museum of Natural History, came very close to identifying Hairy-it as a mammoth, but he never considered checking the spoils pile for the tooth he needed for confirmation. It's also very strange that no geologists returned to the site to check the spoils. The newspaper reporter Waldo Pray heard about the death of Old Bet, but as far as I can tell he never questioned a historian about it. It seems to me that people didn't take seriously the importance of the 1959 discovery to ferret out the full truth. For that, I suppose I can be thankful, because the mystery of these two elephants has led me on the most fascinating search of my life. To know now the true story of each of these remarkable creatures is deeply satisfying. There is of course so much more to learn about Hairy-it that is locked away in her ivory diary, but that is now up to others. After twenty-eight years working at the Maine State Museum, I left in 2001. I hope funding will become available someday to tell the complete story in that slab of tusk at the University of Michigan so that a grand exhibit of Maine's first discovered mammoth and the story of the largest paleontological excavation in the state's history can be viewed by the people of Maine. I feel honored to have been part of such an amazing story and to have been drawn so deeply into the legend of Old Bet that I could not resist searching for the truth about her fascinating life and the impact she had on the birth of the American circus.

ACKNOWLEDGEMENTS

First, I wish to thank Patricia Newell and North Country Press for selecting my manuscript for publication. Writing it during the pandemic isolation was my lifeline to sanity. Acceptance was a great reward.

This book almost didn't get written. I procrastinated over the years after I left my position at the Maine State Museum in 2001 about doing a popular account, then settled into a new life on an island six miles off the Maine coast. But one day my neighbor, Belva Staples, sent me a post on Facebook of an old newspaper report showing me holding a mammoth tooth. That resulted in a talk on the discovery at our local library, then another during Science Week in Bangor, Maine. It was just the kick in the pants that I needed.

Thanks to my next door neighbor, Gary Rainford, who is a fabulous poet and teaches creative writing online for a college in New York state, I developed the skills and confidence to tackle The Search for Elephants in the Maine Woods Project. For over two years I researched, wrote, and illustrated the manuscript.

Hal Borns, Dan Fisher, Ann Marie McGuire, Cathy Harriton, and Jill Trask read the manuscript and made thoughtful suggestions.

I could never have led the scientific endeavor without the encouragement of many people, most notably Hal Borns, J. R. Phillips, Bruce Bourque, and Dan Fisher. Each of these people filled a special need for me. Hal acted as my overall advisor on the project to my great relief. I consulted with Bruce on many technical details and worked periodically with his assistant, Bob Lewis. J. R. dove into the nuts and bolts of the project and kept me on track in regard to my responsibilities. My conversations with Dan Fisher were always illuminating. I could count on his wisdom at every juncture of the project. Later he would go on to work on the frozen megafauna in Siberia.

Ron Harvey was invaluable to my museum colleague Linda Carroll and me as we wrestled with all the considerations of bone recovery and stabilization. The Canadian team of Bob Grantham and Kelley Kosera gave us valuable insight into the problems of wet site excavation and subsequent bone preservation. My experience in the badlands of Montana was a great introduction to bone recovery. Team leader Diane Gabriel provided many opportunities to learn the proper professional field methods for a successful excavation. Sadly Diane died from recurrent breast cancer two years after my return to Maine.

I greatly appreciate the fine work of Linda Carrell and Judy Ritchie, both at the mammoth site and in the conservation laboratory. Linda was exceptional in maintaining the lab after we lost our inhouse conservators. I could always rest easy when I relied on Linda's involvement. Judy proved to be top volunteer in so many ways that I cannot list.

Greg Hart was important in the logistical planning, still and video photography, mechanics, and general construction at the mammoth site.

Scott Mosher was a man for all seasons on the project. As museum workshop foreman, he assisted Greg Hart in planning the temporary infrastructure at the dig site, from ramps and processing equipment to a functioning kitchen and other living arrangements. But he also worked with conservation staff to help streamline their work and assisted in many areas where a competent hand was needed.

The geologists at the site headed up by Tom Weddle from the Maine State Geological Survey were astounding, constantly sampling and debating the importance of the site. I learned a great deal from these people, especially from Chris Dorian and Lisa Churchill who spent considerable time at the site and had many mealtime conversations with me.

I was greatly impressed with the professionalism of reporters who came to the site. In particular I must single out those associated with WCSH-TV. The narrative produced by these people became that year's story of the year for the station.

I was blessed with many dedicated volunteers during the two-year excavation and subsequent bone stabilization process. Several of these have been mentioned in my text, but many are now lost to my fading memory.

In regard to the history of Old Bet, I am indebted to the helpful librarians at the Maine State Library and the Hubbard Free Library in Hallowell, Maine. Brett Mizelle, Professor of History and Director of the American Studies Program at California State University Long Beach, provided me with an invaluable documented exhibition timeline for both the Crowninshield elephant and Old Bet, as well as other helpful material. Early in my research Tina Vickery, doing her own research in the Maine State Library, came across the newspaper article of Esther Moody recalling her mother's happy experience with Old Bet. It became the ideal juxtaposition to the murder of the elephant and the sad commentary on Daniel Davis's life provided to me by Bruce Tucker.

I tend to be a bit of a Luddite. Thankfully, I'm married to a computer wiz who has helped me enormously. Anytime I have encountered a problem in creating this document, Jeanne has always been there to advise and assist me in the structure of my narrative as I strain my brain to remember. Thank you, sweet lady.

The last contributors I must mention were of vital importance to the Hairy-it Project. They are the landowners of the mammoth site. After the excavation and publication of our scientific paper, the landowners lifted their restriction on disclosure. I am thankful to Wallace Fengler and his late wife

Sylvia for allowing this historic excavation to occur. They could see the value of sharing this find with the people of Maine. For that I am sincerely grateful.

CH. 1
How it all began

1. From author's notes. An interview with William J. Littlejohn.
2. *Portland Press Herald*, September 1, 1959.
3. *Portland Evening Express*, September 4, 1959.
4. *Portland Press Herald*, September 5, 1959.

CH. 2
Got Bone?

1. Dibner Award – Maine State Museum, Henry Ford Museum, and the Smithsonian Institution's Museum of American History – Engines of Change. See https://hssonline.org/grants-prizes/shot-dibner-award-for-excellence-in-museum-exhibits-2020

CH. 3
A Provenance Dilemma

1. L. M. Eastman, L. M. "The Portland Society of Natural History: The Rise and Fall of a Venerable Institution," *Northeastern Naturalist*, 13, (Monograph 1, 2006) 1-38.

CH. 6
Extremely Unstable

1. *Portland Press Herald*, September 1, 1959.
2. Maine State Museum Conservation Laboratory Report 12/10/90. Conservation Work Reference Number: CL 90263.

CH. 7
Empirical Proof

1. Since our dating of the tusk other fossils of terrestrial Ice Age animals have been found in Maine, such as musk ox teeth and those of an extinct species of horse. See Andrew M. Barton, *et al.*, *The Changing Nature of the Maine Woods* (U of NH Press, 2012), 41-42.

CH. 8
The Fisher of Ancient Elephants

1. Radiocarbon dating revealed that the whale died in 1926.

In September of 1993 I received a letter from Alan Ruffman, President of Geomarine Associates Ltd., in Halifax, Nova Scotia. One of his requests was for information about a possible tsunami event that had occurred on the Maine coast that I was unaware of.

His letter stated that "the January 9, 1926, possible tsunami was seen as first an emptying then a sudden inrushing of water at about noon at Bass Harbor, Vinalhaven, Corea and Bernard."

Those locations span the neighborhood of Great Cranberry Isle, which is three miles out to sea from Acadia National Park. It is possible that the whale was stranded at that time in the cove where diver Wesley Bracey found the ulna.

CH. 12
Gaining Experience

1. The survey was done in Makoshika State Park. At over 11,500 acres, it is the largest state park in Montana. The name Makoshika is derived from a Lakota phrase meaning bad land or land of bad spirits, alluding to its wasteland appearance and possibly the giant bones of dinosaurs that periodically wash out of the hillsides.

CH. 14
A Little Clam's Importance

1. Andrew M. Barton *et al*, *The Changing Nature of the Maine Woods* (University of New Hampshire Press, 2012), 25.

CH. 16
Hairy-it

1. *Portland Press Herald*, September 5, 1959.

NOTES

CH. 24
A Bone to Pick

1. For those interested in seeing a video of this discovery go to https://www.facebook.com/watch/?v=10155631679551981

CH. 27
The Diary of Hairy-it

1. Zvi Goffer, *Archaeological Chemistry*, (John Wiley and Sons Publishers, 2006), 287.

CH. 28
The Ghost of Old Bet

1. *Portland Press Herald*, (September 10, 1959), 40.

CH. 29
Shunning the Dark Continent

1. https://www.loc.gov/rr/program/bib/ourdocs/jay.html
2. George G. Goodwin quotes a letter from the collection of the Essex Institute in his article, "The Crowninshield Elephant," *Natural History Magazine*, October 1951.
3. See footnote on page 68 of The Travelers' Charleston: Accounts of Charleston and Lowcountry, South Carolina, 1666-1861, edited by Jennie Holton Fant, University of South Carolina Press 2016.

4. "America's First Elephant at Harvard Graduation Exercises in 18th Century Tour of the Continent," (no writer attributed) *Harvard Crimson*, October 6, 1934.

5. Fant, *The Travelers' Charleston*, 67 – 68.

6. Stuart Thayer, "The Elephant in America Before 1840," in the Circus Historical Society's journal, *Bandwagon* 31, no. 1 (1987): 20 – 26. Thayer also included the following information. "All this is told us by the *Independent Chronicle* of Boston dated 28 June."

7. *United States Chronicle* (Providence), June 29, 1797. Endnote no. 25 of "I Have Brought my Pig to a Fine Market" by Brett Mizelle, in *Cultural Change and the Market Revolution*, 1789-1860, edited by Scott C. Martin, 211.

CH. 31

An Elephant, an Artist Showman, and a Wealthy Sea Captain

1. https://www.palaeos.com/systematics/greatchainofbeing/ greatchainofbeing.html

2. *Natural History of Quadrupeds* (Harper & Brothers, 82 Cliff Street 1839) 28-30.

3. Bruce A. Castleman, *Knickerbocker Commodore: The Life and Times of John Drake Sloat 1781-1867*, (State University of New York Press, Albany 2016) 24.

4. George S. Bryan, *The First American Circus*, (*The Mentor World Traveler* vol. 10, No. 3, April 1922), 34. [Unfortunately, Bryan did not list his primary source of information.] Also, "She was brought to America by a sea captain as a speculation: and, according to the best information I have been able to obtain, was purchased by one John Sloat,

who in turn sold her to Hachaliah Bailey,..." ("Old Bet and the Origin of the American Circus," *The Dearborn Independent* vol. 2, February 25, 1922) 2.

5. *Imagining George Washington: The Boglewood Catalogue of Images* (Introduction).

6. https://philadelphiaencyclopedia.org/archive/peales-philadelphia-museum/

7. From the Edward Savage Biography, National Gallery of Art (nga.gov).

8. Harman C. Westervelt, "John Pintard," (*The Chronotype*, January, 1873) 154-155.

9. Terry Ariano, "Beasts and Ballyhoo, The Menagerie Men of Somers" (*Westchester Historian*, Summer 2008). It is also possible that the animal was the Crowninshield Elephant (*Political Barometer*, Poughkeepsie, April 16, 1805).

CH. 32

A Sea Captain, an Elephant, and a Farmer Meet at a Tavern

1. Chauncey Mitchell Depew, "Peekskill Centennial Address," *Speeches and LiteraryContributions at Fourscore and Four* (New York, 1918, Harvard College Library) 113

2. George L. Chindahl, "A Chronology of American Menageries and Circuses," *Bandwagon*, Vol. 2, No. 4 (Jul – Aug), 1958, 9.

3. "Famous Old Tavern on Astor House Site," *Worcester Daily Spy* (Worcester, Massachusetts 28 January 1902) 3. From the "Old Newspapers Tell the History of Two Manhattan Taverns," blog.genealogybank.com.

4. *Knickerbocker Commodore*, 26.

5. Stephen died that year, so he may not have been present during this incident.

6. Stonehouse and Related Families, Person Page – 173.

7. Hachaliah Lyman Bailey, geni.com.

8. https://www.nyhistory.org/community/slavery-end-new-york-state

CH. 34
Dealing With the Elephant in the Room

1. www.stlzoo.org/animals/abouttheanimals/.../asianelephant/elephantedibles

2. George S. Bryan, "Old Bet and the Origin of the American Circus," *The Dearborn Independent* (Dearborn, Michigan, February 25, 1922, Vol. 22), 2. See also Terry Ariano in the Westchester Historian (summer 2008)

3. *Constitutional Democrat*, November 4, 1806. Also, *Charleston Courier*, January 27 and February 7, 1807.

4. *Beasts and Ballyhoo*, 3.

5. *History of Westchester County*, (New York, Vol. 2), 488.

6. This fragment is in the collection of the Somers Historical Society.

CH. 36
A Blast from the Tropic Past Affects an Elephant Lass

1. Today called the East Indies, but more formally known as Nusa Tenggara.

2. Bernice De Jong-Boers, "Environmental History of the Island of Sumbawa (Indonesia)," *IIAS Newsletter, No 10, Regions – Southeast Asia.*

3. Gillen D'Arcy Wood, *Tambora*, (Princeton University Press, 2014), 15.

4. Sophia Raffles, *Memoir of the Life and Public Services of Sir Thomas Raffles*, 242.

5. Wm. and Nicholas Klingaman, "Tambora Erupts in 1815 and Changes World History," *Scientific American*, March 1, 2013.

6. *Raffles*, 247.

7. British Lieutenant Owen Phillips interview with the Raja of Sanggar, *Raffles*, 249.

8. Ibid, 249.

9. "Tambora," 21, *Raffles*, 275-276

10. *Raffles*, 273-274.

11. Todd McLeish, "URI volcanologist discovers lost kingdom of Tambora," *URI Today* (post) Feb. 27, 2006.

12. Robert Evans "Tambora," (*Smithsonian Magazine*, July 2002).

CH. 37

New Eyes on America's Ice Age Prize

1. Dellihay, Tom D.; Ocampo, Carlos (November 18, 2015). "New Archaeological Evidence for an Early Human Presence at Monte Verde, Chile," *PLOS ONE.* **10** (11) : e0141923.

2. Adovasio, J. M., *et al. The Meadowcroft Rockshelter Radiocarbon Chronology.* **55**(2): 348-354.

3. Waters, Michael R.; *et al.* (25 March 2011). "The Buttermilk Creek Complex and the Origin of Clovis at the Debra L. Friedkin Site, Texas," *Science.* **331**(6024) 1599-1603.

4. Holen, Steven R.; *et al.* "A 130,000-year-old archaeological site in southern California, USA," *Nature* **544**(7651): 479-483.

5. Erlandson, Jon M.; *et al.* "The Kelp Highway Hypothesis: Marine Ecology, the Coastal Migration Theory, and the Peopling of the Americas," *The Journal of Island and Coastal Archaeology.* 2(2) 161-174 October 2007.

6. Montaigne, Fen, "The Fertile Shore," *Smithsonian Magazine,* January, 2020.

CH. 38

It's a King Solomon Thing

1. As of this writing, the house and barn still stand at 3 Route 138, Somers [https://www.zillow.com>New York>Somers>10589]. Also, nyshistoricnewspapers.org/lccn/sn83008557/1974.

2. My thanks to Brett Mizelle for finding this bit of information in the *Savannah Republican*, 25 October, 1810.

3. "The Public are respectfully informed, that the ELEPHANT has returned to the city of Albany, there to winter at the same stand that it stood the last winter, viz. at WETMORE'S INN, corner of Beaver and Green streets, where it may be seen every day (Sundays excepted) from 9 o'clock A. M. to 9 P. M. The Elephant is 14 years old, upwards of eight feet high, and weighs more than 6000 pounds. Admittance 25 cents, Children half price. November 10, 1814" (https://www.flickr.com/photos/albanygroup/)

4. www.nextexithistory.com/dowdens-ordinary-the-elephant-comes-to-clarksburg/

5. *The Journal News* (White Plains, New York) 26 Aug. 1984, Sun. page 457. [Newspapers.Com]

6. Wm. Slout, *Olympians of the Sawdust Circle, A Biographical Dictionary of the Nineteenth Century American Circus,* (The Borgo Press, San Bernardino, CA. Copyright 1998)

7. *The Journal News.*

8. Philip A. Loring , "The Most Resilient Show on Earth: The Circus as a Model for Viewing Identity, Change, and Chaos," (Ecology and Society 12(1):9 2007)

CH. 39

A Fragile Steppe for Hairy-it

1. R. Dale Guthrie, "Mammals of the mammoth steppe as paleoenvironmental indicators," In Hopkins D. M., Schweger C. E., Young S. B. (ed.): *Paleoecology of Beringia* (Academic Press, New York, 1982) 307 – 329.

2. R. Dale Guthrie, "Origin and cause of the mammoth steppe," *Quaternary Science Reviews*, 20(2001), 574 – 649.

3. Eske Willerslev *et al,* "Fifty thousand years of Arctic vegetation and megafaunal diet," (*Nature* 506, 47 – 51 16 February 2014).

4. Ross Anderson, "Pleistocene Park," *The Atlantic*, April 2017.

5. Guthrie, Ibid.

6. Sergey A. Zimov *et al*, "The Past and Future of the Mammoth Steppe Ecosystem," 12 March 2012 *Paleontology in Ecology and Conservation*, 193–225.

CH. 40
Frankenstein, the Vampire, and an Elephant's Chilly Walk

1. Diary of Rev. William Bentley, entry Sept. 1, 1797.
2. Ibid.
3. *Biddeford Journal*, May 25,1933.
4. Wednesday, June 12, 1816. *Eastern Argus*.
5. Quoted in the April 5, 1934, *Eastport Sentinel*.
6. Shelly-Godwin Archive shelleygodwinarchive.org

CH. 41
Killing Time on the Mammoth Steppe

1. Charles C. Mann, "The Clovis Point and the Discovery of America's First Culture," *Smithsonian Magazine*, November 2013.

CH. 42
From Bad to Worse

1. An old rhyme quoted by author Lee-Lee Schlegel in her article, "The Year Without a Summer," 1816, In Maine for the Milbridge Historical Society, Milbridge, Maine.
2. The *Eastern Argus* newspaper in Portland, Maine.

3. According to granddaughter Nellie Clark Strong (Somerville, Mass.) noted in "The Year Without a Summer" by Michael Steinberg, *Old Farmer's Almanac* 2018. Also in Lee-Lee Schlegel.

4. *The Hallowell Gazette* in Hallowell, Maine, June 12, 1816.

5. Lee-Lee Schlegel.

6. *The Diary of William Bentley Pastor of the East Church Salem Massachusetts*, Vol. 4 January, 1811 – December, 1819, (The Essex Institute, Salem, Mass. June 12, 1816), 392.

7. Lee-Lee Schlegel.

8. John V. Dippel, *1800 and Froze to Death: The Impact of America's First Climate Crisis.* (Algora Publishing, 2015) 37-38

9. C. A. Stevens, "1800 and Froze to Death," *Our Paper*, Vol. 24, No. 51, Concord Junction, Massachusetts, Dec. 19, 1908.

10. Lee-Lee Schlegel.

CH. 43
The Mammoth Who Lost Her Steppe

1. Harold W. Borns Jr. *et al*, "The deglaciation of Maine," *U. S. A.*, 2004.

2. W. Thompson *et al.* "Late Wisconsinan Glacial Deposits etc.," 58th Field Conference of the Northeast Friends of the Pleistocene.

3. P. G. Holland, (Davis 1965) quoted in "Pleistocene refuge areas and the revegetation of Nova Scotia, Canada", *SAGE Journals* 1981.

4. (Guthrie, 2006) quoted by Eric Post in his 2013 book *Ecology of Climate Change: The Importance of Biotic Interactions*, (Princeton University Press), 40.

CH. 44
Getting to Know Elephant Bet

1. Dexter W. Fellows and Andrew A. Freeman, *This Way to the Big Top: The Life of Dexter Fellows,* (The Viking Press, New York, 1936).

2. A witness to this relationship penned the story to the *National Register*, which was subsequently printed in Hallowell, Maine's *American Advocate* and *Kennebec Advertiser* on Saturday, September 14, 1816. There is no historical record of Nathan Howes's dog's name. I chose a name common to the 19th century.

3. A paraphrase of a portion of no. 2.

CH. 45
Goodbye Hairy-it

1. (Flannery 2001; Putshkov 2001; Mammothsite.com 2004) quoted in "Pleistocene: The Big Chill," Donald Davis, in the 2006 book *Southern United States: An Environmental History,* (Donald Davis *et al.*), 5.

2. John Merck's April 2011 article, "Musings on Alaska During the Last Ice Age" geol.umd.edu

3. Hoyle, B. Gary; *et al. Late Pleistocene mammoth remains from Coastal Maine, USA.* Quaternary Research 61(2004) 277-288.

CH. 46
Old Bet's Final Walk

1. (*Biddeford Journal*, May 25, 1933)

2. See note 3 for Chapter 38.

3. John V. H. Dippel, *Eighteen Hundred and Froze to Death: The Impact of America's First Climate Crisis*, (Algora Publishing 2015) 25.

4. An itinerary for displaying the elephant is found in advertisements in the issues of the *Hallowell Gazette* and the *American Advertiser* of July 1816. Weather events are also described there.

5. *Hallowell Gazette*, Wed., July 17, 1816, Vol. III, No. 29.

6. For an excellent history of "Ohio Fever" see *The Pioneers* by David McCullough, (Simon & Schuster, 2019).

7. John V. H. Dippel, "Scarcity of Crops," *Boston Gazette*, 15 July 1816. Quoted in *Eighteen Hundred and Froze to Death: The Impact of America's First Climate Crisis*, John V. H. Dippel. (Algora Publishing 2015), 25.

8. https://www.lewistonmaine.gov/421/History-of-Lewiston

9. Modern day Auburn.

10. Brooke Nasser , "North Bridge: A passage across the river," *Bangor Daily News* October 30, 2016.

11. Pilings had likely been driven into the riverbed prior to 1812. For details on a traditional ferry run, see William Peberdy, "An Antique Ferry," (*Country Life in America*, September, 1907), 598.

12. This quote is from a 1903 interview with Mrs. Esther Moody recalling the story her mother told about the elephant being exhibited in the Lewiston area in the year Esther was born: 1816. Reprinted in the February 2, 1918, edition of the *Lewiston Saturday Journal*.

CH. 48
Adding to Despair

1. Very little was known about Daniel Davis, Jr. until recent research by Bruce Tucker, President of the Alfred Historical Society. I am indebted to Bruce for allowing me to use his research for this publication. Bruce wrote in a May 16, 2019 e-mail to me the following:

   > Daniel Davis and his brother David Davis were sawmill owners in the northern portion of Alfred called "the Gore". They had borrowed considerable money to expand their milling interests but their plans were sunk by the embargo in the War of 1812. To compound the brother's problems, Daniel's brother David died in Sept 1815, leaving Daniel responsible for the support of David's family and the debt the pair owed. Daniel had, at that point, moved onto David's homestead and sold his own land to fend off creditors. In January 1816, probate courts decreed that David's widow was to have David's assets appraised in October and sold at auction Dec. 4, 1816. This, of course, would leave Daniel landless and still responsible for his brothers family and his own. Daniel's nephews eventually sold the family homestead in the Gore and the clan reportedly relocated to Freedom, Maine.

2. *American Advocate*, Sat. Sept. 14, 1816. Page 1.
3. *The Reminiscences of Neal Dow*, (The Evening Express Publishing Company, Portland, Maine. 1898) 46-47.
4. Ibid.

5. This is known thanks to Lucius Perkins's correspondence with the librarian of Alfred's Parsons Memorial Library in March of 1926.

   Mr. Perkins wrote:

   'I had it from an old gentleman, Usher P. Hall, years ago that some trouble arose relating to admission to the barn where this elephant was being exhibited at the village [Alfred]…The story goes that a man teased the elephant by giving her a piece of tobacco which provoked the owner and the two men had a quarrel.'

6. "She lived about forty minutes after being shot, and retained her docility to the last moment." (Extract of a letter to the editor of the *Boston Gazette*, dated Alfred, York County, 26th July 1816.)

7. *Portland Gazette,* Tuesday, July 30,1816.

8. *Essex Register* of Wednesday, July 31, 1816.

9. *Murder of the Elephant. An Accurate Account of the Death of that Noble Animal the Elephant;* Furnished by a gentleman of Alfred, York County, July 6, 1816. (Printed by Nathaniel Coverly, Milk-Street), 4.

10. *Eastern Argus*, Wednesday, July 31,1816.

CH. 49

A Double Resurrection

1. Extract of a letter to the Editor of the Boston Gazette, dated Alfred, York County, 26th July 1816.

2. Thursday, November 27, 1817, The Evening Post (New York, New York page 3).

3. Columbian Centinial, December 13, 1817.]

CH. 50
End of Another Legend

1. *Ladies Literary Cabinet*, (Nathaniel Smith and Co. New York, New York, August 11, 1821), 112.
2. *New York Post*, December 31, 1821.
3. Elephant Hotel History, somershistoricalsoc.org
4. Somers Hamlet Historic District, Somers Town, Westchester County, Somers, New York 10589.
   Livingplaces.com/NY/Somers/Somers_Hamlet_Historic_District.html
5. Ibid.

CH. 51
Drawing the Final Curtain

1. Joel J. Orosz, *Curators and Culture: The Museum Movement in America, 1740 – 1870*, 132.
2. Joel J. Orosz, Quoted in "Curators and Culture: The Museum Movement in America," 1740 to 1870, (University of Alabama Press), 117.
3. Phineas Taylor Barnum, *The Autobiography of P. T. Barnum* (London: Ward and Lock, 1855), 11.
4. Ibid, page 82.
5. Ibid, page 85.
6. *The Life of P. T. Barnum*, (P. T. Barnum, Buffalo: Courier Co. 1888), 63.
7. Phineas Taylor Barnum, *The Autobiography of P. T. Barnum* (London: Ward and Lock, 1855), 86.
8. Ibid, page 139.

9. Ibid, page 39.

CH. 52
Old Bet Remembered

1. "Plaque To Mark Slaying Of First Circus Elephant," July 19, 1963, *Portland Press Herald*.

ABOUT THE AUTHOR

Most of Gary Hoyle's work has been at the interface of art and science. For 28 years he was an exhibits artist and the Curator of Natural History at the Maine State Museum. His part-time project work has been for institutions from New England to Iowa, including the Field Museum of Natural History in Chicago. His personal artwork has toured the country from coast to coast, as well as Japan. He is now a great grandfather and lives with his wife Jeanne the librarian and Olaf the cat on an island off the coast of Maine. For more information go to his website: garyhoyle.weebly.com.

www.ingramcontent.com/pod-product-compliance
Lightning Source LLC
Chambersburg PA
CBHW041733200326
41518CB00020B/2584